Materials Forming, Machining and Tribology

For further volumes:
http://www.springer.com/series/11181

J. Paulo Davim
Editor

Tribology in Manufacturing Technology

Editor
J. Paulo Davim
Department of Mechanical Engineering
Campus Santiago
University of Aveiro
Aveiro
Portugal

ISSN 2195-0911 ISSN 2195-092X (electronic)
ISBN 978-3-642-31682-1 ISBN 978-3-642-31683-8 (eBook)
DOI 10.1007/978-3-642-31683-8
Springer Heidelberg New York Dordrecht London

Library of Congress Control Number: 2012943973

Printed on acid-free paper

Springer is part of Springer Science+Business Media (www.springer.com)

Preface

Tribology is the science and engineering of *interacting surfaces in relative motion and associated issues*, such as friction, lubrication, and wear. It incorporates various disciplines such as surface physics and chemistry, fluid mechanics, materials, contact mechanics, and lubrication systems.

Tribology has an important role in manufacturing technology because all processes involve surface contact mechanics. Manufacturing processes of many kinds involve tribological considerations. Several examples have been chosen to illustrate how the tribological concepts have been applied to improve the manufacturing technology.

The Chap. 1 of the book provides tribology of cutting tools. Chapter 2 is dedicated to the tribology of machining. Chapter 3 described tribology in metal forming processes. Chapter 4 contains information on tribology in hot rolling of steel strip and Chap. 5 is dedicated of micro contact at interface between tool and and workpiece in metalforming. Finally, Chap. 6 contains information on coatings and applications.

The present book can be used as a research book for final undergraduate engineering course or as a topic on manufacturing at the postgraduate level. Also, this book can serve as a useful reference for academics, tribology, and manufacturing researchers, manufacturing, materials, and mechanical engineers, professional in manufacturing, and related industries. The interest of scientific in this book is evident for many important centers of the research, laboratories, and universities as well as industry. Therefore, it is hoped this book will inspire and enthuse others to undertake research in this field of tribology in manufacturing technology.

The Editor acknowledges Springer for this opportunity and for their enthusiastic and professional support. Finally, I would like to thank all the chapter authors for their availability for this work.

Aveiro, Portugal, August 2012 J. Paulo Davim

Contents

Contributors

Viktor P. Astakhov 1792 Elk Ln, Okemos, MI 48864, USA, e-mail: astakhov@scientist.com

Akira Azushima Department of Mechanical Enginering, Yokohama National University, 79-5 Tokiwadai Hodogaya-ku, Yokohama 204-8501, Japan, e-mail: azu@ynu.ac.jp

Sérgio Tonini Button Department of Materials Engineering, School of Mechanical Engineering, University of Campinas, Campinas, Brazil, e-mail: sergio1@fem.unicamp.br

J. P. Davim Department of Mechanical Engineering, University of Aveiro, Campus Santiago, Aveiro 3810-193, Portugal, e-mail: pdavim@ua.pt

Z. Y. Jiang School of Mechanical, Materials and Mechatronic Engineering, University of Wollongong, Wollongong, NSW 2522, Australia, e-mail: jiang@uow.edu.au

M. J. Jackson Center for Advanced Manufacturing, Purdue University, West Lafayette, IN, USA, e-mail: bondedabrasive@gmail.com

J. Morrel Y-12 National Security Complex, Oak Ridge, TN, USA, e-mail: morrelljs@y12.doe.gov

C. A. S. de Oliveira Mechanical Engineering Department, Universidade Federal de Santa Catarina, Florianópolis, SC, Brazil, e-mail: carlosa@emc.ufsc.br

G. M. Robinson Center for Advanced Manufacturing, Purdue University, West Lafayette, IN, USA

Cássio A. Suski Instituto Federal Catarinense, 3122 St, n. 340, apt. 901, Balneário Camboriú, SC 88330-290, Brazil, e-mail: cassio_suski@hotmail.com

M. Whitfield Center for Advanced Manufacturing, Purdue University, West Lafayette, IN, USA

D. B. Wei School of Mechanical, Materials and Mechatronic Engineering, University of Wollongong, Wollongong, NSW 2522, Australia, e-mail: dwei@uow.edu.au

Chapter 1
Tribology of Cutting Tools

Viktor P. Astakhov

Abstract This chapter introduces the concept of the cutting tool tribology of a part of the metal cutting tribology. It argues that the importance of the subject became actual only recently because the modern level of the components of the machining system can fully support improvements in the cutting tool tribology. The underlying principle for tribological consideration is pointed out. The major parameters of the tribological interfaces, namely the tool-chip and tool-workpiece interfaces are considered. The chapter also describes an emerging mechanism of tool wear known as cobalt leaching. The rest of the chapter presents some improvements of tribological conditions of cutting tools as the use of application specific grades of tool materials, advanced coating and high-pressure metal working fluid (MWF) supply.

1.1 What is Tribology of Cutting Tools?

1.1.1 Tribology of Metal Cutting

The term tribology comes from the Greek word *tribos*, meaning friction, and *logos*, meaning law. Tribology is therefore defined as "*the science and technology of interactive surfaces moving in relation to each other.*" The science of *Tribology* concentrates on contact physics and the mechanics of moving interfaces that generally involve energy dissipation. Its findings are primarily applicable in mechanical engineering and design where tribological interfaces are used to transmit, distribute

V. P. Astakhov (✉)
General Motors Business Unit of PSMi, 1792 Elk Ln, Okemos, MI 48864, USA
e-mail: astakhov@scientist.com

J. P. Davim (ed.), *Tribology in Manufacturing Technology*, Materials Forming, Machining and Tribology, DOI: 10.1007/978-3-642-31683-8_1,

and/or convert energy. The contact between two materials, and the friction that one exercises on the other, causes an inevitable process of wear. What those contact conditions are, how to strengthen the resistance of contact surfaces to the resulting wear, as well as optimizing the power transmitted by mechanical systems and complex lubrication they require, have become a specialized applied science and technical discipline which has seen major growth in recent decades. Bearing the rather colorful name "Applied tribology," this field of research and application encompasses the scientific fields of contact mechanics, kinematics, applied physics, surface topology, hydro- and thermodynamics, and many other engineering fields under a common umbrella, related to a great variety of physical and chemical processes and reactions that occur at tribological interfaces.

When it comes to metal cutting, tribology is thought of as *something* that has to be studied in order to reduce the tool wear (and thus increase tool life). Although this is true in general, it does not exhaust the application of tribological knowledge in metal cutting. Unfortunately, the published books and articles on the subject do not treat the subject in a systematic way. Rather, the collection of non-correlated facts on tool materials, cutting regimes, tool life and its assessment, cutting fluids, tool coatings, etc., is considered as the tribology of metal cutting. Having read the known works and related materials, one does not feel thoroughly equipped to analyze and improve the tribological conditions in various metal cutting operations. This is because of the commonly understood meaning of "metal cutting tribology," which is something related to reduction of tool wear, its assessment and reduction. Although it is true that cutting tool wear and its proper assessment is a part of metal cutting tribology, the assessment and reduction of tool wear are only "natural by-products" of this field of study.

Astakhov has pointed out that the ultimate objective of metal cutting tribology is the optimization of the cutting process through minimizing the energy spent in metal cutting by considering all the physical and chemical processes that take place simultaneously in this process [1]. Increased tool life, improved integrity of the machined surface, higher process efficiency and stability are the results of achieving this goal. The starting point of such optimization is the consideration of energy partition in the cutting used for assessment of physical efficiency of the cutting system. The latter is defined as a ratio of the energy needed to separate of the layer being removed from the rest of the workpiece and the total energy required by the cutting system for its existence provided that the nominator in such a ratio is minimized.

1.1.2 Tribology of Cutting Tools

The tribology of cutting tools is defined as a part of metal cutting tribology that relates to the cutting tool itself, and thus used for its proper design and performance. Figure 1.1 shows the graphical representation of the cutting tool tribosystem. Geometry, mechanics and physic of such a system are systemically related

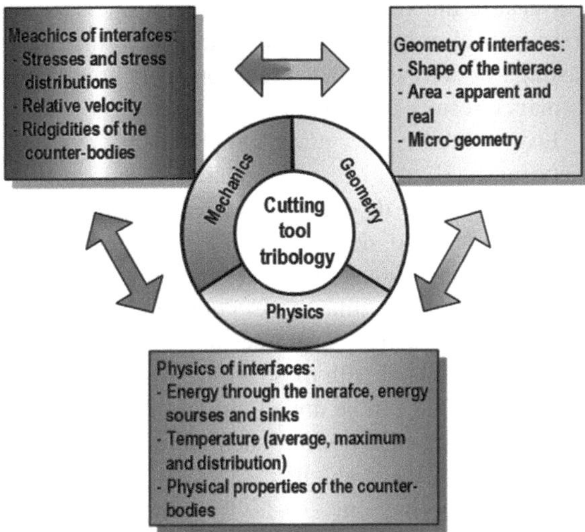

Fig. 1.1 Graphical representation of cutting tool tribological system

so that it is improper to consider each of these groups of processes individually. For example, if one changes geometry of the tool rake face by altering (restricting) the contact length and/or rake angle then maximum stress and its distribution over the tool chip interface change that, in turn, may change the contact temperature. As a result, the nature of tribological processes taking place on this interface may change significantly that, in turn, may affect the cutting tool, tool wear pattern and tool life.

The main objective of the cutting tool tribology is to optimize the tribological conditions at the tool–chip and tool–workpiece interfaces. The understanding of these conditions and the proper utilization of this understanding in the design of cutting tools should result in the increased overall efficiency of the cutting system and in the reduction of tool premature failures which is a crucial issue in modern unattended production lines and manufacturing cells. These results can be used in the meaningful selection of the machining regime, tool design including its geometry and tool materials (including coatings).

1.1.3 Importance of the Subject

Recent survey indicates that in the automotive and mold-making industries:

- The correct cutting tool geometry is selected less than 30 % of the time.
- The tool is used at the rated cutting regime only 48 % of the time.
- Only 57 % of the tools are used up to their full tool-life capability.
- The correct tool material is selected less than 30 % of the time.
- The correct cutting fluid (coolant) parameters are used 42 % of the time.

Because the USA spends approximately $150 billion annularly to perform its conventional metal cutting operation [2] that range from simple clean-up of castings or welds to high precision work, the cost of subpar metal cutting performance results in overwhelming losses.

Low reliability of cutting tools and sporadic tool failures in advanced manufacturing facilities (i.e., in the automotive industry) are the major hurdles in the way of wide use of efficient unattended machining production lines and manufacturing cells to decrease the direct labor costs and improve efficiency of machining operations. As such, significant downtime due to low reliability of drilling tools undermines the potential of high-efficient production lines and manufacturing cells rising question about feasibility of unattended machining operations. To deal with these issues, a closer look at the cutting tool is needed to understand what can be done to improve the situation. In the author's opinion, the most promising direction for such an improvement is the tribological approach to the analysis of the problem.

In metal cutting, only 30–50 % of the energy required by the cutting system is spent for the useful work, i.e. for the separation of the layer from the workpiece [1]. This means that 25–60 % of the energy consumed by the cutting system is simply wasted. A great part of this wasted energy is spent at the tool–chip and tool–workpiece interfaces due to non-optimized tribological processes. This wasted energy has the following effects at different levels:

- Plant/machine shop level: Increases power consumption of the whole machine shop/manufacturing plant, causing needs to use high power supply cables, transformers, switches etc.
- Machine tools: Increases the resultant cutting force and torque causing a need to build more powerful machines and their components
- Cutting process: Increases contact temperatures, lowers tool life, affects the shape of the produced chip, and leads to the necessity of using different cooling media that, in turn, lowers the efficiency of the machining system as more energy is needed for cooling medium delivery and maintenance.

The optimization of tribological processes at the tool–chip and tool–workpiece interfaces results in the following:

- *Reduction of the energy spent in cutting*. Because the efficiency of the cutting system is low due to energy losses during tribological interactions, the optimization of the tribological processes improves the efficiency of the cutting system by reducing the energy spent by the cutting system.
- *Proper selection of application-specific tool material (coating)*. Considering the energy transmitted through the tribological interfaces in metal cutting, one can select a tool material for a given application to assure the chosen performance criterion such as tool life, quality of the machined surface, efficiency, etc.
- *Proper selection of tool geometry*. It becomes possible as the cutting tool tribology correlates the parameters of tool geometry, as a whole, and the parameters of tool microgeometry, in particular, with tribological processes at the tool–chip

and tool–workpiece interfaces, and thus the optimized tribological parameters can be directly used in the selection of proper tool geometry.
- *Control of machining residual stresses in the machined surfaces*. The machining residual stresses are determined by the tribological process taking place at both the tool–chip and tool–workpiece interfaces. Therefore, one can control both the superficial and in-depth machining residual stresses over a wide range in terms of their sign, magnitude and distribution by controlling the tribological processes at the mentioned interfaces.
- *Proper selection of cooling and lubricating media as well as the method of its delivery and application technique*. In modern high- speed machining, the metal cutting fluid (MCF) is normally supplied through the cutting tool so the proper selection and application of a particular medium is only possible when true tribological mechanism of its action is known.

1.1.4 Why Now?

This is an interesting frequently asked question. Although many components of the cutting tool tribology have been studying since the earlier years of the 20th century, the implementation of the results of these studies was rather modest. The most apparent cause for this is that these studies lacked systemic approach, i.e. one component, for example tool life, was studied while other important parameters, e.g. process efficiency, were left unconsidered. However, in the author's opinion, the real cause is that neither the whole machining systems, as a whole, nor its component, the cutting tool, were ready for implementation of any advances made in tribological studies.

In the not-too-distant past, the components of the machining system were far from ideal in terms of assuring normal tool performance, and thus could not get any advantage of applying optimization of tribological process at the tool contact interfaces. Tool specialists (design, manufacturing, and application) were frustrated with old machine tools having spindles that could be rocked by hands; part fixtures that clamped parts differently every time; part materials with inclusions and great scatter in the essential properties; tool holders that could not hold tools without excessive runouts assuring their proper position; starting bushing and bushing plates that had been used for years without replacement; low-concentration often contaminated MCFs which brought more damage than benefits to the cutting tool; manual sharpening of cutting tools eyeballing the accuracy of such a grinding; available on machines cutting speed and feed with limited range; low dynamic rigidity of machines; etc. As such, the most advanced (and thus expensive) cutting tools performed practically the same (or even worse) as the basic tools made in a local tool shop. As a result, any further development in the tool improvement including its tribological optimization was discouraged as leading tool manufacturers did not see any return on the investments in such developments.

For many years, a stable though fragile balance between low-quality (relatively inexpensive tools) and poor machining system characteristics was maintained. Metal cutting research was attributed mainly to university labs and their results were mostly of academic interest rather than of practical significance. It is clear that the metal cutting theory and the cutting tool tribology advancements designs based on this theory were not requested by practice as practical specialists have not observed any difference in the performance of a tribologically optimized cutting tool compare to that made in a nearby tool shop.

This has been rapidly changing since the beginning of the 21st century as global competition forced many manufacturing companies, first of all car manufacturers, to increase efficiency and quality of machining operations. To address these issues, leading tool and machine manufacturers have developed a number of new products—new tool materials and coatings, new cutting inserts and tool designs, new tool holders, powerful precision machines, part fixtures, advanced controllers, which provide a wide spectrum of information on cutting processes and other aspects of machining and increase the efficiency of machining operations in industry by increasing working speeds, feed rates, tool life, and reliability. The major changes are as follows:

Machine Tools

Dramatic changes in the design and manufacturing of machine tool can be summarized as follows:

1. Machines with powerful digitally-controlled truly-high-speed motor-spindles. For example, machine with working rotational speed of 25,000 rpm and 35 kW motor-spindles are used in the advanced manufacturing powertrain facilities in the automotive industry. New facilities are designed to use 35,000 rpm machine with 40 kW motor-spindles.
2. New spindles assure tool runout <0.5 μm. High static and dynamic rigidity of such spindles and machines made with granite beds assures chatter (vibration)-free performance even at most heavy cuts at truly high-speed machining conditions.
3. New machines are equipped with high-pressure (approx. 70 bars) MCF supply through the cutting tools to provide cooling and lubrication needed for high-speed operations. MCF cleaned to 5 μm are delivered at constant controlled temperatures suitable for a given machining operation.

Tool Holders and Tool Pre-Setting Practice

A tool holder has two interfaces. One of them is the interface with the machine tool while the other—with the cutting tool. Tool holder developed for high-speed machining have HSK interface with the machine tool and shrink-fit interface with the cutting tool.

The HSK tool holder is designed to provide simultaneous fit on both the spindle face and the spindle taper. Supporting the cutting tool and holder in both the axial and radial planes creates a significantly more rigid connection between the tool and spindle.

The shrink fit technology has been developed for high-speed machining to minimize tool runout and to maximize tool holding rigidity. With shrink fit toolholders vibration is reduced and cutting is noticeably faster and smoother due in part to the lack of set screws and component tolerance variances. Shrink-fit toolholders assure 50,000 N holding force, tool runout for normal tool holders is 5 μm max and 3 μm for precision tool holders.

For years, tool pre-setting was one of the weakest links in assuring tool proper position and performance. No matter how good are the tool and its tool holder, tool pre-setting accuracy determines the actual tool performance. Needs for more accurate tool pre-setting, tool data transformability from the pre-setter to the machine tool controller, and tool performance traceability have led to the development of non-contact laser automated tool pre-setting machines. Nowadays, tool pre-setting machines (Zollar and Kelch, for example) are widely used in high-speed machining applications.

Each tool contains an ID number to enable users to retrieve and use this data later on. The tool is loaded into the holder, and the CNC-driven tool length stop is said to ensure correct tool positioning—for example, tool tip to gage line. Measurement results are transferred to the tool identification chip (the Balluff chip) embedded in the holders. Such pre-setting machine can provide accuracy to within 3 μ on each tool, which, in turn, results in improved machining quality.

Advanced Cutting Process Monitoring

Many recent technologies offer tool and machine monitoring, from detecting whether an intact tool is present to measuring a tool's profile. Some can even measure the power consumed by the spindle motor and use that information to control the feed rate and minimize machining time. The most common features of modern machine tool controllers developed for high-speed unattended manufacturing are:

- *Detecting broken or absent tools*. Small, simple tool detectors that check for the presence of a drill or other cylindrical-shaft tool were developed and implemented on modern machine tools. Their small size and simple operation adapt them well for many environments, including machining centers, screw machines and transfer machines.
- *Power monitoring*. Directly monitoring the power consumed by the spindle motor allows one to understand exactly what is happening with the tool. A new, sharp tool requires less power to cut than a worn, dull tool. Power monitoring systems are available that take their data directly from the motor controller; others measure with transducers on the wiring to the motor. It is called 'sensory perception' for machines—The system obtains information that would not otherwise be detectable and notifies the operator as soon as possible when there is a problem.
- *Adaptation*. Not only can power-monitoring units detect broken or worn tools, but with an "adaptive control" option, it can also use them to control the feed rate, reducing machining time, yet extending tool life by keeping the tool load constant and well under maximum stress.

Advances in Cutting Tool Materials

Modern grades of powder-metallurgy high-speed steels combined with advanced coatings introduced for high cutting speeds that previously were considered suitable only for carbide tools. Today, it is common to use dry hobbing with high-speed steel hobs at 80 m/min. Modern grades of polycrystalline diamond (PCD) tool material allow milling, drilling, and reaming of high-silicon aluminum alloys at speeds of 1,000–11,000 m/min. Modern grades of carbide tools combined with advanced coatings allow machining of alloyed steels at speeds of 300–600 m/min. Modern grades of PCBN (polycrystalline cubic boron nitride) allowed to introduce hard machining operation which substituted grinding operations. New tool materials and advanced grades of the existing tool materials, including nano-coatings, were introduced.

Advances in the Cutting Tool Manufacturing

There are many advances in the cutting tool manufacturing are taking place rapidly. Among them, the introduction of CNC tool grinders/sharpeners and tool geometry measuring machines are probably most significant.

For decades, manual tool grinding/sharpening machines were used in the cutting tool industry. It was no possible to maintain the geometry of ground/sharpened tools with reasonable accuracy which, moreover, varied significantly from one re-sharpening to the next. An exact tool geometry simulated by any advanced tool design program could not be reproduced by such machine. It was also not possible to grind any complicated profile of the tool as it might be necessary in order to optimize tool performance. Naturally, any advanced tool geometry suitable for optimal performance of a machining operation was bluntly rejected by machining practice as been "impractical" for a real world application.

Such a situation with tool grinding has been changing rapidly since the beginning of 21st century. Today's tool and cutter grinder is typically a CNC machine tool, usually 4, 5, or 6 axes, which produces end mills, drills, etc. which are widely used in the metal cutting industry. High levels of automation, as well as automatic in-machine tool measurement and compensation, allow extended periods of unmanned production. With careful process configuration and appropriate tool support, tolerances less than 5 μm can be consistently achieved even on the most complex parts.

No matter how good is the fully optimized cutting tool geometry (using, for example, a FEM simulation software) and how well it is depicted in multiple section planes on the tool drawing made using a 3D CAD program, it practically useless if such an optimized geometry cannot be reproduced and then inspected with high accuracy. Until very recently, the most common practice of measuring tool geometry was manual inspection with not very accurate subjective results that depend on the inspector experience, tool complexity and many other factors. Naturally, the accuracy of such an inspection was not nearly sufficient for assurance of effective performance of the cutting tool.

To address this important issue, Zollar and Watlters developed CNC tool inspection machines capable of accurate inspection of cutting tools. For example, ZOLLER-genius 3 measuring and inspection machine is equipped with 5 CNC-controlled axes for measurement and inspection of virtually any tool parameter—

fast, simple, precise to the micron and fully automatic. Equipped with a 500-fold magnification incident light camera, Genius 3 can automatically inspect micro tools down to 0.1 mm. The machine includes measuring programs for virtually every parameter (radius contour tracing, effective cutting angle, clearance angle, helical pitch and angle, chafer width, groove depth, tumble and concentricity compensation, step measurement, etc.).

The foregoing analysis suggests that many components of modern metal machining systems are ready to adopt the advances that can be made in the cutting tool tribology. This book section aims to address some important issues in the cutting tool tribology.

1.2 Underlying Principle

The underlying principles for the cutting tool tribology considerations constitute a unique viewpoint from which the subject is considered. They unify various facets of such considerations establishing physically-sound criteria at each stage. The underlying principles include the definition of the metal cutting process and the basic laws. The basic laws in any branch of applied science establish its foundation principles. For example, the laws of thermodynamics are amongst the simplest, most elegant, and most impressive products of modern science. Most physical laws are designed to explain processes which humans experience in nature. The best and the most general way to establish such laws is to consider the energy flow and transformation in a technical system distinguished by a given science because only system considerations result in physics-based and sound laws.

The process of metal cutting is defined as a forming process, which takes place in the components of the cutting system that are so arranged that the external energy applied to the cutting system causes the purposeful fracture of the layer being removed. This fracture occurs due to the combined stress including the continuously changing bending stress causing a cyclic nature of this process. The most important property in metal cutting studies is the system time. The system time was introduced as a new variable in the analysis of the metal cutting system and it was conclusively proven that the relevant properties of the cutting system's components are time dependent. The dynamic interactions of these components take place in the cutting process, causing a cyclic nature of this process [3].

Consequences:

It follows from this definition that, considered together (the system approach), the following features distinguish metal cutting among other closely related manufacturing processes and operations:

- *Bending moment*. The bending moment forms the combined stress in the deformation zone that significantly reduces the resistance of the work material to cutting. As a result, metal cutting is the most energy efficient material removal process (energy per removed volume accounting for the achieved accuracy) compare to other closely related operations.

- *Purposeful (micro) fracture of the layer being removed under combined stress.* The fracture occurs in each successive cycle of chip formation. Stress triaxiality significantly in the deformation zone affects the fracture strain, and consequently the energy required for plastic deformation of the layer being removed is also affected. This observation is significant because: (a) the energy associated with this deformation constitutes up to 70 % of the total energy required by the cutting system for its existence [4], and (b) this plastic deformation is nuisance of metal cutting so that the associated energy is a total waste [1], the reduction of this energy by adjusting stress triaxiality is the major reserve in increasing efficiency of the cutting process. This adjustment can be made by the tool geometry parameters, selection of the machining regime and other particularities of the cutting process.
- *Stress singularity at the cutting edge.* The maximum combined stress does not act at the cutting edge compared to other closely related forming operations. Rather, a (micro) crack forms in front of the cutting edge. As a result, when the cutting system is rigid and the cutting tool is made and run properly, the wear occurs at a certain distance from the cutting edge that allows maintaining the accuracy of machining over the entire time of tool life.
- *Cyclical nature.* Metal cutting is inherently a cyclic process. As such, a single chip fragment forms in each chip formation cycle. As a result, considered at the appropriate magnification, the chip structure is not uniform. Rather, it consists of chip fragments and connectors. The frequency of the chip formation process (known also as the chip segmentation frequency) primarily depends on the cutting speed and on the work material. The cutting feed and the depth of cut (>1 mm) have very small influence on this frequency.

As the above-provided definition of the metal cutting process is system based, the second law of metal cutting [1] a.k.a. the deformation law is formulated as the follows: *Plastic deformation of the layer being removed in its transformation into the chip is the greatest nuisance in metal cutting, i.e., while it is needed to accomplish the process, it does not add any value to the finished part. Therefore, being by far the greatest part of the total energy required by the cutting system, the energy spent on this deformation must be considered as a waste which should be minimized to achieve higher process efficiency.*

It directly follows from the second law of metal cutting that the greater the energy of plastic deformation, the lower the tool life, quality of the machined surface, and process efficiency. Therefore, the prime objective of the cutting process design is to reduce this energy to its lowest possible minimum by the proper selection of the tool geometry, tool material, machining regime, MCF, and other design and process parameters. In other words, any change in the cutting (or even, machining) system is considered beneficial if it results in reducing this energy under the maximum temperature and process efficiency constrains.

Because simple and accurate method to assess the amount of plastic deformation in metal cutting has been developed and verified that allows optimization of the metal cutting process and tool design even at the shop floor level [5], real machining can be used for verification of the optimization results.

Fig. 1.2 The tool-chip interface as viewed in machining (Sandvik Coromant Co)

1.3 Tribological Interfaces: Tool-Chip Interface

1.3.1 Geometry of the Tool-Chip Interface

When a metal is cut, the cutting force acts mainly through a small area of the rake face, which is in contact with the chip and thus, is known as the tool–chip interface. Therefore, it is of interest in determining the cutting force, developing the theory of tool wear and understanding the mechanics of chip formation to establish the tribological characteristics of the tool–chip interface. The first step to achieve this goal is to define the geometry of this interface properly as many contradictive models are presented in the literature on the matter.

Figure 1.2 shows the tool-chip interface as viewed in machining (Sandvik Coromant Co). As can be seen, the tool cuts the chip from the workpiece. After being cut, the chip first follows the tool rake face and then it curves away. Figure 1.3 shows a model of real machining shown in Fig. 1.2 including the basic terminology used in the further considerations. As can be seen, the depth of cut, a_p and the tool cutting edge angle, κ_r determine the chip width as

$$b_1 = a_p / \sin \kappa_r \tag{1.1}$$

Note that this depth determines the width of the tool-chip interface [1, 6, 7].

Figure 1.4 shows View I in Fig. 1.3 as related to the tool-chip interface where the tool-chip contact length is designated as l_c. Multiple experimental studies conducted by Zorev [8], Loladze [9], Poletica [10] and many others conclusively

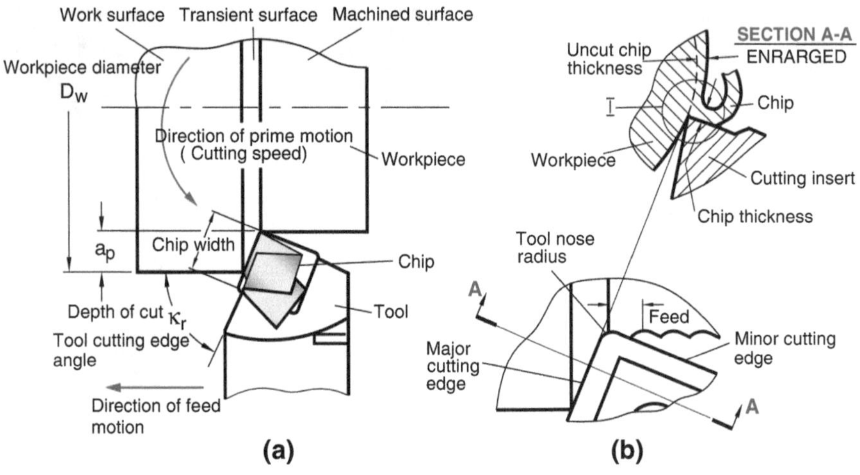

Fig. 1.3 Model of real machining shown in Fig. 1.2 including basic terminology

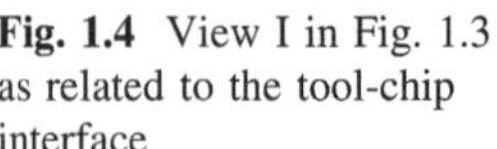

Fig. 1.4 View I in Fig. 1.3 as related to the tool-chip interface

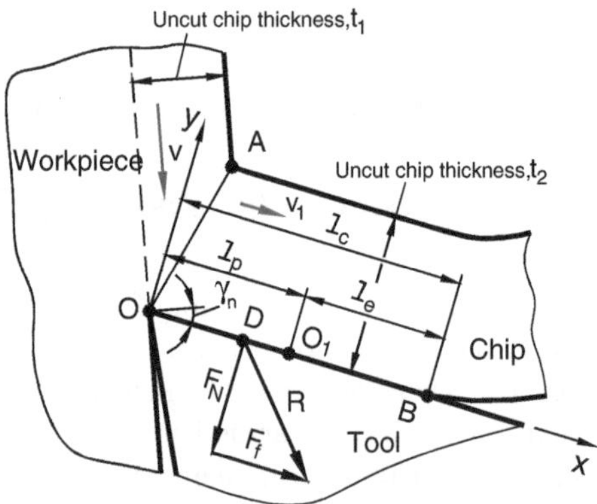

proved that the tool-chip interface consists of the parts, namely plastic and elastic and therefore in the model shown in Fig. 1.4, the whole contact length l_c is divided into two distinctive parts: the plastic part, l_{c-p} which extends from the cutting edge (point O) to a certain point O_1 and the elastic part, l_{c-e} from point O_1 to the point of tool-chip separation B.

Summarizing the results of multiple experiments, Abuladze [11] proposed the following expression to calculate to the length of the plastic part of the tool-chip interface

$$l_{c-p} = t_1[\zeta(1 - \tan\gamma_n) + \sec\gamma_n] \tag{1.2}$$

where t_1 is the uncut chip thickness determined through the cutting feed f and the tool cutting edge angle, κ_r as $t_1 = f \sin\kappa_r$, γ_n is the normal/orthogonal (depending on the particular tool geometry) rake angle (the rake angle in Sect. 1.1 in Fig. 1.3), and ζ is the chip compression ratio (hereafter CCR) defined as $\zeta = t_2/t_1$ where t_2 is the actual chip thickness.

Astakhov introduced a similarity criterion to be used in metal cutting tribology, named the Poletica criterion [1]

$$Po = \frac{l_c}{t_1} \quad (1.3)$$

The Poletica criterion, *Po*-criterion, strongly depends on CCR and weakly depends on the rake angle. Moreover, it was found that *Po*-criterion remains invariant to changes in the mechanical and physical properties of the work material, and thus can be used to determine the chip-contact length as follows

$$Po = \frac{l_c}{t_1} = \zeta^{k_t} \quad (1.4)$$

where $k_t = 1.5$when $\zeta_a < 4$and $k_t = 1.3$when $\zeta_a \geq 4$.

The correlation between the Poletica criterion and CCR obtained experimentally and expressed by Eq. (1.4) represents the condition of static equilibrium of the chip [1].

1.3.2 Summary of Tribological Conditions at the Tool-Chip Interface

Determining the tribological conditions at the tool–chip interface qualitatively allows one to select the proper tool material, coating, geometry, method of MCF supply and other important practical parameters normally considered in the tool design/selection process. Unfortunately today, such parameters available in the literature are either qualitative or too specific as obtained for a particular set of cutting parameters, i.e. the available information cannot provide any practical help to the tool/process designer in the selection of cutting tool parameters. This section aims to discuss tribological conditions at the tool–chip interface in the quantitative yet general manner useful for practical application as well as for cutting tool research/optimization.

The basic tribological characteristics of the tool–chip interface are:

1. The contact length–the length of the tool–chip contact, *l_c* It can be determined using Eq. (1.4).
2. The relative velocity of the counter bodies at the tool–chip interface, i.e. the chip sliding velocity with respect to the tool rake face (Fig. 1.4), $v_1 = v/\zeta$.
3. The friction force at the tool–chip interface as acting on the tool, F_f.
4. The specific frictional force which is the mean shear stress,$\tau_c = F_f/(l_c b_1)$. Distribution of the shear stress over the contact length $\tau_c = f(l_c)$.
5. The normal force at the tool–chip interface acting on the tool, F_N.

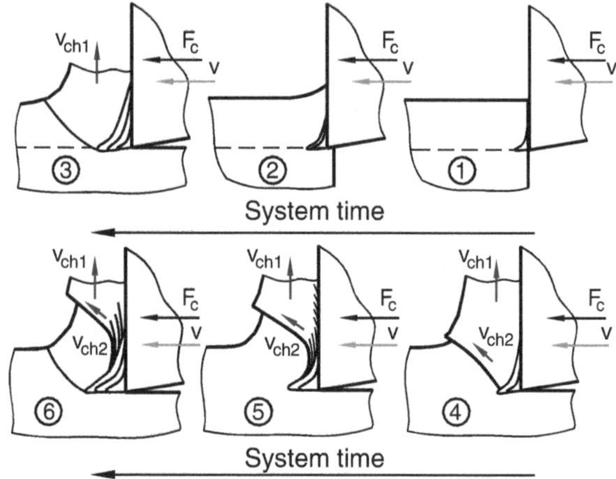

Fig. 1.5 System consideration of the chip formation process in metal cutting

6. Mean normal stress at the tool-chip interface, $\sigma_c = F_N/(l_c b_1)$. Distribution of the normal stress over the contact length $\sigma_c = f(l_c)$.
7. Mean contact temperature at the tool–chip interface,

1.3.3 System Consideration

The major difference between the current and all other tribological studies/ recommendations dealing with the cutting tool is the system approach discussed in Sect. 1.2. In the author's opinion, non-systemic considerations of the cutting process and tribological phenomena involved have caused a wide variety of the results on stress and temperature distribution over the tool contact interfaces (discussed earlier by the author [1]). Therefore, before proceeding any further, a basic of system approach, i.e. the nature of the cyclic character of the cutting process is considered in this section.

Figure 1.5 shows the system consideration of the metal cutting model [1]. Phase 1 is the initial stage. When the tool is in contact with the workpiece, the application of the cutting force F_c leads to the formation of a deformation zone ahead of the cutting edge. The tool moves forward with a cutting speed v. The workpiece first deforms elastically and then plastically. As a result, an elastoplastic zone forms ahead of the tool that allows the tool to advance further into the workpiece so that a part of the layer being removed comes into close contact with the tool rake face (phase 2). When full contact is achieved, the state of stress ahead of the tool becomes complex including a combination of bending and compressive stresses. The dimensions of the deformation zone and the maximum stress increase with the cutting force F_c. When the combined stress in this zone reaches a limit (for a given work material), a sliding surface forms in the direction of the maximum combined stress (phase 3). The partially formed chip starts to slide with a velocity v_{ch1} relative to the tool rake face. This instant may be considered as the very beginning of chip formation. As soon as the sliding surface forms, all the chip-cantilever material starts to slide along this surface with a velocity v_{ch2} while the whole chip slides with a velocity v_{ch1} along the rake face (phase 4). Upon sliding, the resistance to the tool penetration decreases, leading to a decrease in the dimensions of the plastic part of the deformation zone. However, the structure of the work material, which has been deformed plastically and now returns to the elastic state, is different from that of the original material. Its appearance corresponds to the structure of the cold-worked material. Experimental studies [3, 8, 11–13] showed that the hardness of this material is much higher than that of the original material. The results of an experimental study using a computer-triggered, quick-stop device proved that this material spread over the tool–chip interface by the moving chip constitutes the well-known chip contact layer (phase 5), which is now believed to be formed owing to severe friction conditions in the so-called secondary deformation zone that exists over the tool-chip interface [8].

The chip fragment continues to slide until the force acting on this fragment from the tool is reduced, when a new portion of the work material enters into the contact with the tool rake face. This new portion attracts part of the cutting force F_c. As a result, the stress along the sliding surface diminishes, becoming less than the limiting stress and arresting the sliding. A new fragment of the chip starts to form (phase 6).

Chip formed in this way is referred to as the continuous fragmentary chip [12] and it is the most common chip type formed in metal cutting although it appearance may vary depending on particular machining conditions. As such, the cutting speed has major influence [1]. It has a saw-toothed free side and a non-uniform strength along its length. The shear strength of the fragments is much greater than that of fragment connections.

1.3.4 Stress Distribution and Mean

A comprehensive analysis of the distribution of the normal and shear stresses over the tool-chip interface reported in the literature was presented by the author earlier [1]. It was discussed that a great variety of such distribution was reported as obtained using various experimental techniques. The significant differences in the reported results can easily be explained if a system consideration of the metal cutting is used.

Figure 1.6 shows dynamic analysis of the shear and normal stress distribution over the tool-chip interface. As can be seen, the distribution of the shear stress shown in Fig. 1.6a suggests the existence of two zones (plastic and elastic) as discussed above.

Figure 1.6b shows the evolution of the normal stress up to the beginning of each chip formation cycle. As seen, the presence of a partially formed chip affects the normal stress distribution. The maximum normal stress becomes higher when

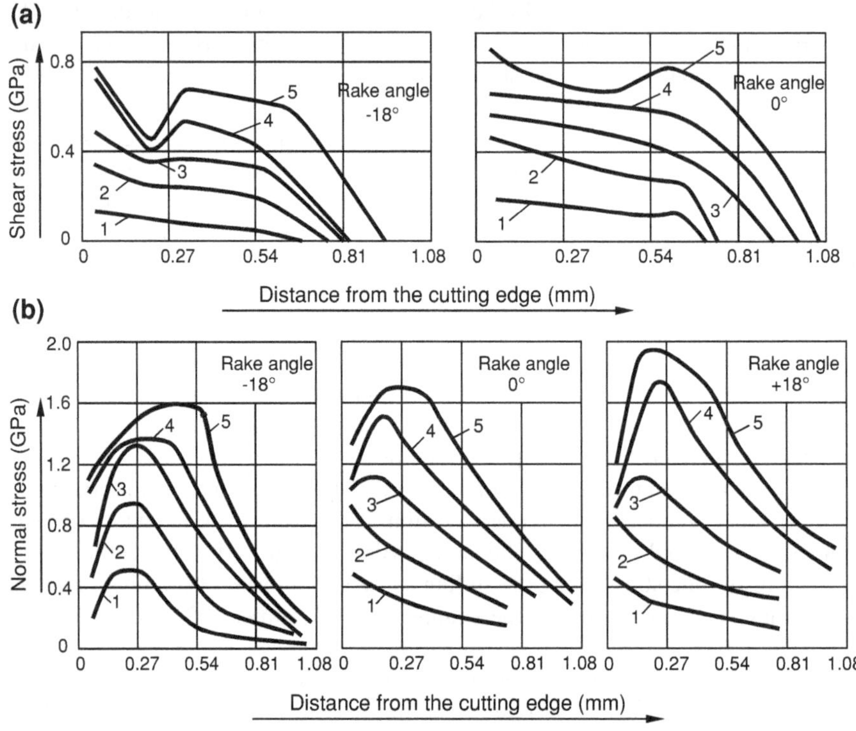

Fig. 1.6 Dynamics shear (tangential) (**a**) and normal (**b**) stress distributions at the beginning of a new chip formation cycle. Curves 1–5 correspond to increasing load with the increment of 250 N, respectively. Work material—AISI 1045 steel (HB180), tool material—carbide P20

increasing the rake angle. Moreover, the location of this maximum shifts towards the cutting edge.

A comparison of the results shown in Fig. 1.6 with the known results [1] shows that there is no contradiction in the reported results. In other words, all the reported normal stress distributions at the tool–chip interface may occur as being considered at different instants over a chip formation cycle and under different cutting conditions.

The above analysis of the normal and shear stress distributions at the tool–chip interface results in two important conclusions:

1. Increasing the rake angle leads to the reduction in the contact length that results in an increase in the maximum shear and normal stresses at this interface. Moreover, the maximum of the normal stress shifts towards the cutting edge. Because the use of high positive tool geometry (which stands for high rake angles) has become a new tendency recently introduced and followed widely by the leading tool producers, the transfer rupture stress of tool materials that represent cutting tool material toughness (at least, to the first approximation) for such applications should be increased.

The application of sub- and nano-grain carbides, for example, made it possible to use cutting tool with high rake angles
2. The fluctuations of the normal and shear stresses at high frequency [3] explain the high scatter in the results of using different coatings. Obviously, these fluctuations should be accounted for in the design of any particular coating.

Analyzing numerous experimental results, Poletica concluded [10] that, although the mean shear stress at the tool chip interface can be correlated with many mechanical properties of the work material, the best fit seems to be achieved with the ultimate tensile strength, σ_{uts}. He concluded that the following empirical relation shows good correlation

$$\tau_c = 0.28\sigma_{UTS} \tag{1.5}$$

The independence of the mean shear stress at tool-chip interface on many factors that affect the cutting process is an important characteristic of this process. Among other factors, the most surprising and seemingly paradoxical is independence of this stress on the mean contact temperature at the tool-chip interface. Zorev [8] and partially Spaans [14] attempted to explain this paradox by mutual influence of two reverse-proportional factors, namely the strain rate and temperature. The following explanation has been offered: the lowering of the mean shear stress at tool-chip interface with the contact temperature is fully compensated by the growth of this stress due to the corresponding increase in the strain rate. In other words, the effect of the temperature on the mean shear stress at tool-chip interface is balanced by the strain rate effect in such a way that this stress remains constant.

This idea, however, were criticized by Poletica [10] who conclusively proved that this is not the case in metal cutting.

The total friction force on the tool rake face (Fig. 1.4) then can be calculated as

$$F_f = \tau_c l_c b_1 \tag{1.6}$$

or

$$F_f = 0.28\sigma_{UTS} t_1 b_1 \zeta^{k_t} \tag{1.7}$$

where b_1 is the chip width (Fig. 1.4). This width is determined based on the concepts of the equivalent cutting edge and chip flow direction as [6]

$$b_{1T} = \sqrt{a_p^2 + \left(a_p \cot\kappa_{r1} + f\right)^2} \tag{1.8}$$

Apart from the mean shear stress at the tool-chip interface, the mean normal stress is highly dependant on the parameters of the cutting process. As discussed by Astakhov [1], the mean normal stress at the tool-chip interface increases with the cutting speed for a wide range of metallic work materials. It also increases with the hardness of the work material and decreases with the rake angle. Summarizing the obtained experimental results, Astakhov [1] showed that not only the uncut chip thickness, cutting speed but also the mechanical properties of the work material affect the mean normal stress insomuch as they affect the Poletica criterion. As this criterion decreases, the mean normal stress increases. The mean contact stress, therefore, is a function (and a characteristic) of the state of stress in the contact zone. It depends on the Poletica criterion in the same way as this criterion affects the state of stress in the contact zone. Function $\sigma_c = f(Po)$can be expressed as

$$\sigma_c = \frac{180}{(Po)^{0.95}} \tag{1.9}$$

which, in the general case, can be represented as

$$\sigma_c = \frac{A_\gamma}{(Po)^{m_\gamma}} \tag{1.10}$$

where coefficient A_γ and power m_γ are determined by the rake angle. As such, normally $m_\gamma < 1$ and A_γ decreases with the rake angle.

The total friction force on the tool rake face (Fig. 1.4) then can be calculated as

$$F_N = A_\gamma t_1 d_w \zeta^{k_t(1-m_\gamma)} \tag{1.11}$$

Because m_γ is close to 1, the power of CCR is very small. Therefore, for a given uncut chip cross-sectional area ($t_1 b_1$), the normal force on the tool rake face is primary a function of the tool rake angle and only weakly depends on CCR.

1.3.5 Temperature

1.3.5.1 Known Facts

Although it is pointed out in almost any book on metal cutting that temperature, and particularly, its distribution has a great influence in machining [15], no one study known to the author quantifies this influence. Instead, it is stated in very general and qualitative terms that temperatures in metal cutting affect "the shear properties" of the work material and, therefore, they affect the chip-forming process itself, and through their effect on the tool, they determine the limits of the process and mode of tool wear. To address each of these points, a great number of works on temperatures in metal cutting have been published. Apart from many contradictive results that can be readily found in the published works and can be logically explained by the difference in the experimental methodologies and accuracy of calibration, numerical and analytical models and the assumptions adopted in both the models, the major concern with these works is their practical significance. In other words, there is no answer to a simple question: "What should one do with the found temperature and its distribution?" The same question arises in any FEM of metal cutting as a common result of such a modeling is colorful field of temperature distribution in the tool, workpiece and chip. A question: "Is the obtained result is good or not?" cannot be answered. In other words, there is no gage to judge 'goodness' or 'optimality' of the obtained temperature results.

Trent and Wright concluded [16] that the major objective of heat consideration in metal cutting is to explain the role of heat in limiting the rate of metal removal when cutting the higher melting point metals. They concluded that there is no direct relationship between cutting forces or power consumptions and the temperature near the cutting edge.

Zorev [8] did not consider temperature as an important factor itself. Considering the energy balance in metal cutting, he calculated that the maximum temperature at the end of the chip formation zone does not exceed 270 °C for plain and alloyed steels while a considerable reduction in the mechanical properties of these materials starts only at temperatures over 300 °C. Therefore, he concluded that metal cutting is a cold-working process where temperature does not affect mechanical properties of the work material in the defamation zone although the chip leaving the cutting tool can be of cherry-red color.

According to Childs et al. [17], the two goals of temperature measurements in machining are: (a) the quantitative measurements of the temperature distribution over the cutting region is more ambitious, but very difficult to achieve, and (b) is less ambitious to measure the average temperature at the tool–chip contact. In the author's opinion, the less ambitious goal makes sense if one knows how to measure this average temperature and, that is more important, how to use the obtained result.

To understand the formation of the temperature fields in the tool, workpiece and the chip, the known publications consider the energy balance (in modern terminology—energy partition) in metal cutting. As the conservation law states and many specialists in metal cutting surprisingly agree with this law (not always the case in metal cutting studies where some fundamental physical laws can be easily declared as inapplicable, for example the principle of minimum energy as discussed by the author's earlier [3, 18]), the almost all the energy required by the cutting system for its existence (referred in the literature as the energy supplied to the cutting system) converts into the thermal energy or simply heat. Small portions of energy stored in the deformed chip and in the cold-worked machined surface hardly exceed 2–3 % of the total energy. Therefore, the power that converts into heat in the cutting system can be calculated rather accurately as $F_c v$, where F_c is the power components of the cutting force and v is the cutting speed.

The next issue is the distribution (partition) of this power (converted in the form of heat) in the cutting system. The heat distribution in the cutting system is originated from study by Schmidt and Roubik [19], who, according to Komanduri [20], carried out calorimetric study in catting and their measurements, thus obtained, permit computation of work, power, forces, average temperature of the chip, etc. They also showed a good agreement between the calorimetric measurements and the power data obtained from torque and thrust measurements.

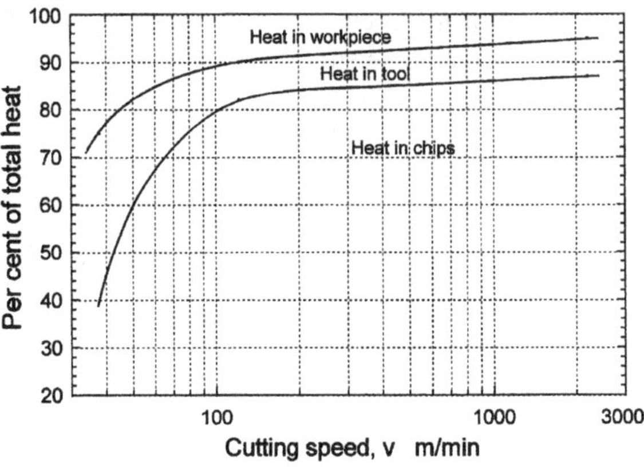

Fig. 1.7 Typical distribution of heat in the workpiece, the tool, and the chips with cutting speed; after Schmidt and Roubik [21]

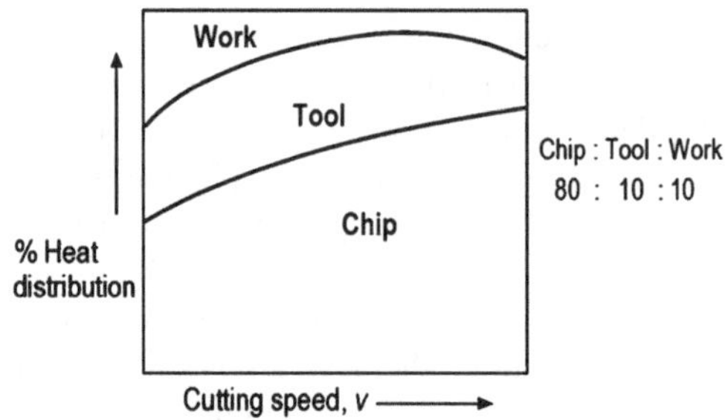

Fig. 1.8 Heat distribution between the chip, workpiece and tool

An example of Schmidt and Roubik results [21] used in the literature is shown in Fig. 1.7. This example and its derivatives have been using in the literature since then (for example, [22, 23]) up to modern book on the subject (for example, [24, 25]). In some modern books, however, this distribution simplified up to that shown in Fig. 1.8 [26], i.e. became of more qualitative than quantitative nature. Our critical analysis of the published data on heat partition in the cutting system revealed an obvious drawback. The partition of heat is always shown as a function of the cutting speed. In other words, the cutting feed, thermal properties of the work and tool materials, influence of MWF and many other 'thermal' particularities of a given machining operation are not accounted for. For example, it is obvious that if a tool material of high thermoconductivity, for example PCD, is used than more heat flows into the tool compare to the case when a tool material of extremely low thermoconductivity, for example ceramics, is used. Therefore, it may be stated that heat partition in metal cutting is application specific and the ratio of the amounts of heat that go into the components of the cutting system is not fixed as shown in Fig. 1.8 but may vary significantly depending upon particularities of a given machining operation.

The common analysis of heat distribution and temperatures in the cutting system is based on the analysis of heat sources. Because practically all of the mechanical energy associated with chip formation ends up as thermal energy [3, 8, 27], the heat balance equation is of prime concern in metal cutting studies. This equation can be written as [3]

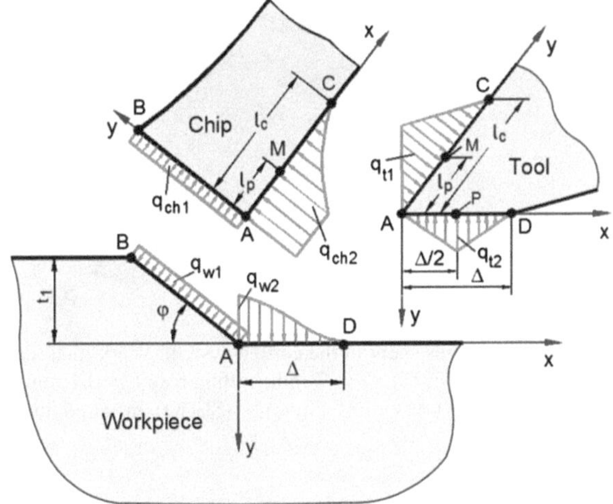

Fig. 1.9 Areas of heat generation on the tool, workpiece and chip

$$F_c v = Q_\Sigma = Q_c + Q_w + Q_t \tag{1.12}$$

where Q_Σ is the total thermal energy (heat) generated in the cutting process, Q_c is the thermal energy transported by the chip, Q_w is the thermal energy conducted into the workpiece, Q_t is the thermal energy conducted into the tool. As shown in Figs. 1.7 and 1.8, under 'normal' cutting conditions, most of the thermal energy generated in the cutting process is conducted into the chip [3, 8, 27].

Example of energy balance shown in Table 1.1, [28] reveals two essential features:

- Most of the thermal energy generated in the cutting process is carried away by the moving chip (80–85 %).
- The higher the cutting speed, the greater portion of the total heat is carried out by the chip.

These facts, however, do not follow from the traditional model of metal cutting. The model shown in Fig. 1.9 [28] illustrates the heat sources on each component of the cutting system, namely, on the tool, workpiece and chip. In this figure, t_1 is the uncut chip thickness, φ is the shear angle, AB is the length of the shear plane, AC is the tool-chip contact length, l_c, AM is the length of the plastic part, l_p of the tool-chip contact length, l_c, AD is the tool-workpiece contact length, Δ.

The thermal energy in the cutting system generates:

1. Due to plastic deformation of the work material on the shear plane, Q_{pd}. This energy partitions into portion that goes to the workpiece $Q_{pd-w} = \int_A^B q_{w1}(y)\cos\varphi dy$ and that goes to the chip $Q_{pd-ch} = \int_A^B q_{ch1}(y)dy$.

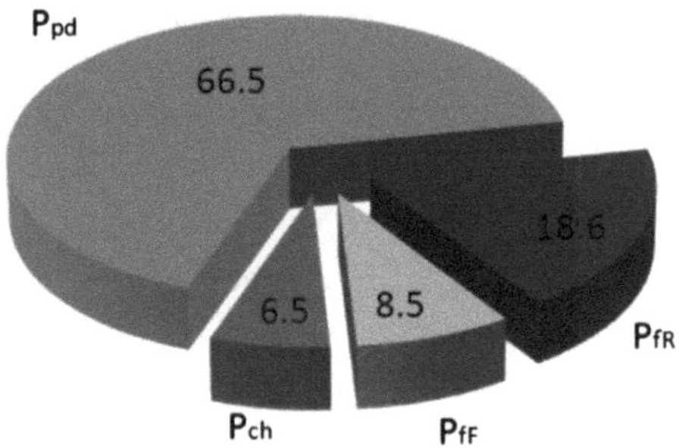

Fig. 1.10 Energies spent in the cutting system. Work material: AISI steel E52100, cutting speed $v = 1$ m/s, depth of cut $d_w = 3$ mm, cutting feed $f = 0.4$ mm/rev; Tool—standard inserts SNMG 432-MF2 TP2500 Materials Group 4 (SECO) installed into a tool holder 453–120141 R1-1 (Sandvik)

2. Due to friction on the tool-chip interface, Q_{Rr}. Its portion $Q_{fR-ch} = \int_A^C q_{ch2}(x)dx$ goes to the chip and $Q_{fR-t} = \int_A^C q_{t1}(y)dy$ goes to the tool.
3. Due to friction on the tool-workpiece interface, Q_{fF}. It portion $Q_{fF-t} = \int_A^D q_{t2}(x)dx$ goes to the tool and that$Q_{fF-w} = \int_A^D q_{w2}(x)dx$goes to the workpiece.

The next question is about the intensity of the heat sources. As discussed in the literature (for example [1, 24, 27], the greatest portion of energy spent in the cutting system is due to plastic deformation of the work material. Figure 1.10 shows an example [4]. In this figure P_{pd} is the energy spent on the plastic deformation of the layer being removed, P_{fR} is the energy spent due to friction at the tool-chip interface, P_{fF} is the energy spent due to friction at the tool-workpiece interface, P_{ch} is the cohesive energy spend on the formation of new surfaces (which can be thought of as spend on the shear plane). As follows, the energy spent on the shear plane is $P_{fR} + P_{ch} = 73$ %. Therefore, 73 % of the total thermal energy generated in cutting is due to plastic deformation of the work material.

As mentioned above, this total energy due to plastic deformation ($P_{fR} + P_{ch}$) is then partitions between the workpiece (portion $Q_{pd\text{-}w}$) and the chip ($Q_{pd\text{-}ch}$). Such a partition, however, does not apparently obey the second law of thermodynamics. The problem is explained as follows (Fig. 1.10

1.3.5.2 Contradiction

Figures 1.11 and 1.12 show the results of actual temperature measurements in the cutting system obtained by Shaw [22] and Astakhov [1]. Similar results were

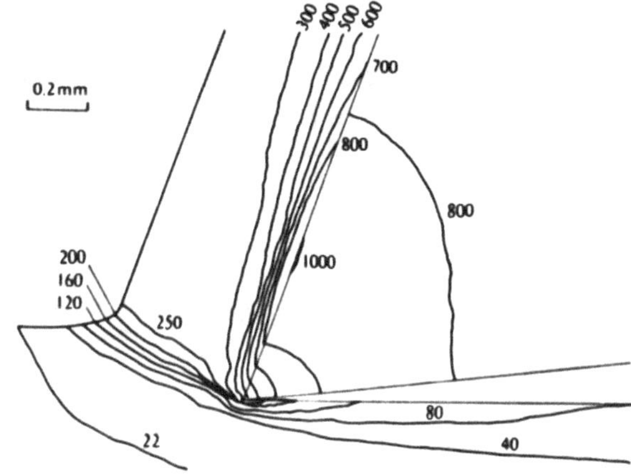

Fig. 1.11 Typical temperature field in metal cutting: Isotherms for dry orthogonal cutting of free machined steel with a carbide tool at cutting speed of 155 m/min and cutting feed of 0.274 mm/rev [22]

obtained by many specialists, for example by Smart and Trent [29], who actually measured not modeled temperature distribution using FEM with unjustifiable input parameters. The comparison of these results with the data shown in Fig. 1.7 and Fig. 1.8 and presented in Table 3.1 accounting for the common model shown in Fig. 1.9, reveals a contradiction with the second law of thermodynamics. This law stated that heat flows naturally from a region of higher temperature to one of lower temperature. Therefore, according to the second law of thermodynamics, portion $Q_{pd\text{-}w}$ should be much higher than $Q_{pd\text{-}ch}$. Experimental results on heat partition, however, shows otherwise, i.e. a way greater part of the total heat flows into the hot small chip than that in the cold large workpiece. This is the discussed contradiction. This contradiction cannot be resolved in principle using the existing notions in metal cutting due to the fact that the traditional model shown in Fig. 1.9 is incorrect [18]. Although the author in his publication tried to explain this revealed contradiction, it seems the explanations were not sufficiently clear.

1.3.5.3 Moving Chip: The Governing Equation

Many cases considered in the literature deal with the so-called stationary systems. There are examples of materials processes in which a solid body is moving out of a hot region and it sheds heat to the environment as it moves away from that heat source. Some examples of this configuration include a long slab of steel emerging from a furnace, a polymer strand leaving an extruder, metal wire being drawn, or a metal rod undergoing continuous induction hardening. The same can be said about moving chip. In many cases, the heat transfer can be approximated as occurring in one dimension (the direction of motion, or the axial direction) and treating heat losses in perpendicular directions as heat sinks. In order for this approximation to be valid, the heat flow in the body must be oriented so that it is mainly in the axial direction. If the heat flux in the direction of the moving body is much greater than the direction normal to motion, then the one-dimensional approximation is reasonable.

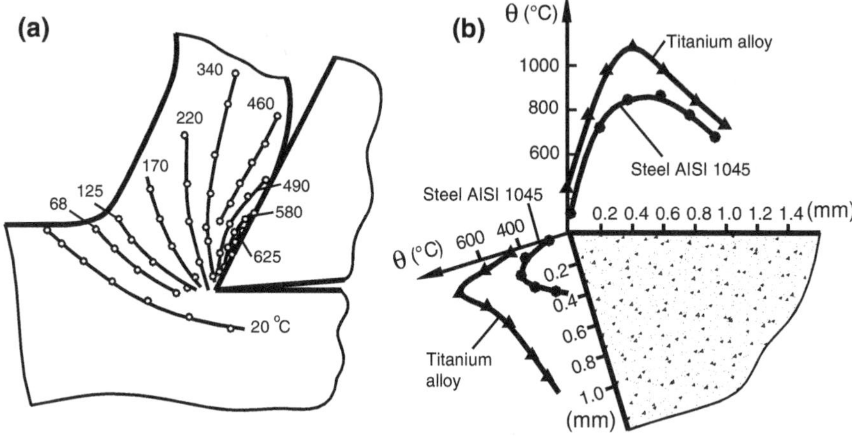

Fig. 1.12 Typical temperature field in metal cutting: **a** Isotherms for dry orthogonal cutting of ANSI 1045 steel with a carbide (P10) tool (rake angle 12°) at cutting speed of 60 m/min and uncut chip thickness 2 mm, **b** Temperature distributions over the tool rake and flank faces. Turning, a carbide cutting tool carbide M20 (92 % WC, 8 % Co), depth of cut $a_p = 1.5$ mm cutting speeds in machining of steel 1045—240 m/min, titanium alloy (Ti6Al4 V)—160 m/min, cutting feed—0.25 mm/rev [1]

Table 3.1 Energy balance in machining (steel 1045)

v(m/s)	Q_{ch} (J/s)	$Q_{ch}/Q_{\sum}$ (%)	Q_w (J/s)	$Q_w/Q_{\sum}$ (%)	Q_{ct} (%)	$Q_{ct}/Q_{\sum}$ (%)	$Q_{\sum}$ (%)
0.10	47.9	50.2	38.4	40.2	9.2	9.6	95.5
0.20	93.7	55.7	63.7	37.8	11.0	66.6	168.4
0.5	272.3	70.3	100.3	25.9	14.7	3.8	287.3
1.00	501.6	76.2	136.9	20.8	19.7	3.0	658.3
2.00	1177.1	82.8	217.5	15.3	27.0	1.0	1421.6
4.00	2306.2	86.3	336.7	12.6	29.4	1.1	2572.3

If the moving body can be modeled as one-dimensional, then one can define a control volume over which he/she can perform an energy balance in order to derive a conservation equation for thermal energy in terms of temperature [30]. In this control volume (of length Δx, cross-sectional area A_{ch}, and perimeter p_v), thermal energy is transferred by conduction (q_x) and advection. Advection is the transport of energy due to the flow of the solid in the x direction through the control volume. The amount of energy which is brought into the control volume at location x by bulk solid motion is ($\dot{m}e_x$), where e_x is the specific enthalpy at x, The mass flow rate (which is constant along the length of the moving body) is $\dot{m} = \rho_{ch} A_{ch} v_{ch}$, where ρ_{ch} is the density of the work material, v_{ch} is the velocity of the chip relative the tool rake face. The rate at which energy is advected out of the volume can be different and is written as ($\dot{m}e_x + \Delta x$). Also, heat can be generated in the volume ($\dot{q}$) and it is also lost to the ambient by convection. For the rest of this derivation, it is assumed that the volume velocity, material properties, and geometry which do not change along the direction of motion (x). It is reasonable assumptions for the chip because as it forms, its velocity relative to the tool and geometry do not change.

Using these conditions, Bejan [30] derived the energy conservation equation which describes the temperature along the length of the moving body, subject to heat generation and convective heat loss in the following form

$$kA_{ch}\frac{d^2T}{dx^2} - \left(\dot{m}c_p\right)\frac{dT}{dx} - hp\left(\theta_{ch} - \theta_{en}\right) + \dot{q}A_{ch} = 0 \qquad (1.13)$$

Axial conduction	Advection	Convection loss	Heat generation

where k is the thermoconductivity of the work material (or material of the chip), c_p is the specific heat of this material, and h_{cv} is the convection heat transfer coefficient of the process, θ_{ch} and θ_{ev} are the temperatures of the chip and environment, respectively.

It is useful to look carefully at this energy equation to remind ourselves of the physical phenomena which govern it. One must never view such an equation in a purely mathematical light, but must keep in mind the physics represented by it. The first term represents the diffusion of thermal energy along the length of the body due to a temperature gradient within it. This diffusion of heat happens regardless of the magnitude of the motion and is independent of it. The second term is the change in the thermal energy of a mass as it moves through space. It is the difference between the energy advected into and out of a control volume of length dx. The third term is the heat lost through convection to the environment and the final term is heat generated in the body. In metal cutting, the chip move very fast so that the convection term can be neglected [1].

The heat transfer by conduction and convention are normally considered in the literature on metal cutting while that by advection does not prevent so much attention. Thermal (or heat) advection is the transport of sensible or latent heat by a moving body, such as the chip in the considered case. Therefore, the role of heat advection, known also as mass transportation, as applicable to metal cutting should be explained.

To do this, Eq. (1.13) is considered together with a simplified model of chip formation shown in Fig. 1.13. In this model, the deformation of the layer being removed into the chip takes place ‘instantly’ on passing the shear plane so that the whole amount of heat due to the plastic deformation is generated along this plane. Being generated, the heat due to plastic deformation may go to the chip due to advection and to the layer being removed due to thermoconductivity. Note that the structure of Eq. (1.13) clearly shows that the generated heat cannot go into the chip by thermoconductivity as per the second law of thermodynamics, i.e. because the temperature of the chip is higher than that of the shear plane and heat goes from a region of higher temperature to that of lower temperature. Therefore, there are two competing mechanisms of heat conduction: thermoconductivity that attempts to

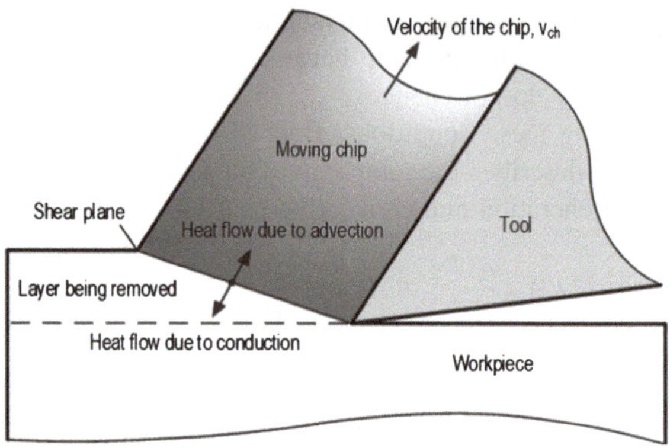

Fig. 1.13 Simplified model of chip formation in metal cutting

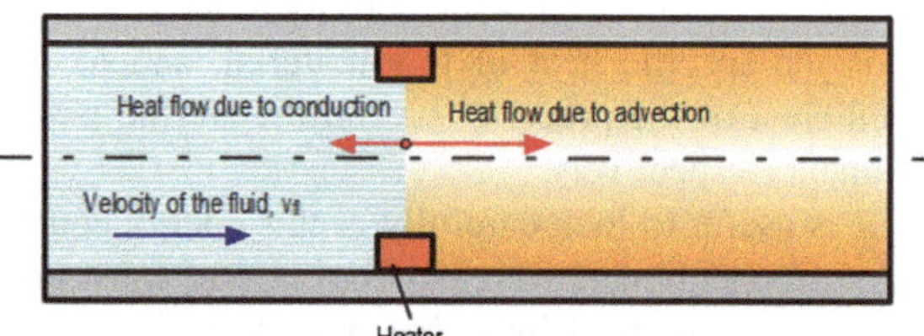

Fig. 1.14 Example of use of the Péclet number

bring a portion of the generated heat into the layer being removed and advertion that attempts to bring a portion of this heat into the chip due to its motion.

The next question to be answered is about the ratio of the portions of the heat generated on the shear plane due to thermoconductivity and that due to advertion. It is well-known in heat transfer studies that such a ratio is determined by the Péclet number [30]. This number is a dimensionless number relevant in the study of transport phenomena in fluid flows. It is named after the French physicist Jean Claude Eugène Péclet. It is defined to be the ratio of the rate of advection of a physical quantity by the flow to the rate of diffusion of the same quantity driven by an appropriate gradient, i.e.

$$Pe \equiv \frac{[\text{advection of heat}]}{[\text{conduction of heat}]} = \frac{VL}{\omega} \tag{1.14}$$

where V is the velocity scale, L is the length scale, and is the thermal diffusivity.

To comprehend the significance of this number, let's consider an example. Figure 1.14 shows a flow of a fluid in a tube where a heater is installed. When the fluid is motionless, i.e. its velocity $v_{fl} = 0$, then the Péclet number is also zero according to its definition given by Eq. (1.14). As such, there is no advertion. The heat from the heater flows in both sides at the same rate. When, however, the fluid velocity becomes $v_{fl} > 0$, then heat advertion takes place so that, according to Eq. (1.13), the temperature on the right side of the heater becomes greater than that on its left side. When the fluid velocity becomes great enough that the Péclet number is equal to ten, then only 1/10 of the heat supplied by the heater flows into the fluid in the left side of the heater while 9/10 of this heat flows to the fluid on its right side. No matter how powerful is the heater, this proportion is still the same.

In metal cutting, the Péclet criterion is represented in terms of machining process parameters as follows [1]

$$Pe = \frac{v_{ch} t_1}{w_w} \tag{1.15}$$

where v_{ch} is the velocity of a moving heat source, i.e. the velocity of chip relative the tool rake face (m/s), w_w is the thermal diffusivity of the work material (m^2/s),

$$w_w = \frac{k_w}{(c_p \rho)_w} \tag{1.16}$$

where k_w is the thermoconductivity of the work material, (J/(m s °C)), $(c_p \cdot \rho)_w$ is the volume specific heat of work material, (J/(m^3 °C)).

As an example, consider machining of AISI 1040 steel under the typical machining conditions: operation—turning; Tool—MTJNR-1616H-09 (ISO 5608:1995) with a carbide insert; cutting speed $v = 3$ m/s (180 m/min); cutting feed $f = 0.25$ mm/rev, chip compression ratio $\zeta = 2$, and thus the velocity of the chip with respect to the tool rake face is calculated as $v_{ch} = v/\zeta = 3/2 = 1.5$ m/s. Thermal diffusivity of the work material is 6.67 10^{-6} m^2/s. For the J-style tool holder, the tool cutting edge angle is $\kappa_r = 93^o$, thus the uncut chip thickness calculates as [7] $t_1 = f \cdot \cos(\kappa_r - 90^o) = 0.25 \cdot \cos(93^o - 90^o) = 0.24965 \approx 0.25$ mm. Thus, the Péclet criterion is calculated as $Pe = (1.5 \cdot 0.25 \cdot 10^{-3})/6.67 \cdot 10^{-6} = 66$. Therefore, 98.5 % of the heat generated on the shear plane due to plastic deformation of the layer being removed flows into the chip while only 1.5 % of this heat flows into the workpiece.

The obtained result has the following significance:

1. It explains the experimentally obtained low temperatures in the workpiece below the shear plane, for example those shown in Figs. 1.11 and 1.12. It explains why at low cutting speed the distribution of heat becomes more even. For example, referring to Table 3.1, when $v = 0.1$ m/s then the amount of heat that goes into the chip is 47.9 % while 38.4 % goes into the workspace.
2. It explains the above-stated contradiction between the experimentally obtained heat balance in metal cutting (Figs. 1.7, 1.8 and Table 3.1) and the model shown in Fig. 1.9. Moreover, it signifies the necessity of the system consideration of the chip formation process in the manner shown in Fig. 1.5 instead of its static analogue exclusively used in the literature.
3. It fully supports statement of Zorev [8] and definition of the cutting process by Astakhov [3] that the metal cutting process is a cold-working process because the temperature of the layer being removed just ahead of the tool hardly exceed 200 °C. In other words, the heat due to plastic deformation of the layer being removed does not affect the mechanical properties of the work material as this heat goes mostly into the chip due to mass transportation, i.e. advection.

One may argue, however, that the shear plane is not a plane in reality but as suggested by some researches, e.g. Spaans and Oxley [14], [31], is a narrow zone. To discuss the influence of temperature in this case, Rosenberg and Rosenberg [31] proposed to estimate the period of time necessary for a microvolume of work material to pass through the deformation zone. It follows from the above discussion that a microvolume of the layer being cut, passing through the shear zone, changes its velocity from the cutting speed v to the chip velocity $v_{ch} = v/\zeta$. Thus, the average velocity of the microvolume is 0.5 v (1–ζ). Therefore, the time T_p necessary to pass the shear zone having the width of h_{sz} would

$$T_p = \frac{h_{sz}}{0.5v\left(1 + \frac{1}{\zeta}\right)} \tag{1.17}$$

Following a suggestion by Spaans, the width of the shear zone is $h = 0.5\ t_1$ [14], one can estimate the time which is necessary for a microvolume to pass the deformation zone for a typical cutting regime using Eq. (1.17). When the workpiece is made of a plain carbon steel, a typical cutting regime is as follows: $v = 120$ m/min $= 2$ m/s; $\zeta = 2.5$; $t_1 = 0.2$ mm. Thus, the estimated time is $T = 0.000071$ s. When the workpiece is made of a high-strength, low-alloy steel, the typical cutting regime may be as: $v = 120$ m/min $= 2$ m/s; $\zeta = 1.3$; $t_1 = 0.05$ mm. As such, $T = 0.000014$ s. As seen, the time necessary for a microvolume to pass the deformation zone is extremely short. As a result, heat generated in this zone due to plastic deformation of the layer being removed can be considered as occurring instantly, i.e. over the shear plane.

1.3.5.4 Summary of Temperature Consideration

The known results of temperature considerations in metal cutting as related to the tribology of cutting tools can be summarized as follows:

1. The tool-chip interface consists of the moving chip and stationary tool. As such, the material of the chip in this contact is highly strain-hardened work material (Fig. 1.5) of high temperature due to plastic deformation of the layer being removed. As shown in Fig. 1.5, the plastically deformed layer (the chip contact layer) has non-uniform thickness being the largest in the region of the cutting edge. This created impression of the so-called secondary deformation zone.
2. The temperature of the chip in not uniforms. The highest temperature is in the highly deformed contact layer while the so-called chip free surface has much lower temperature. When the uncut (undeformed) chip thickness is large enough and or when the chip compression ratio is small, the temperature of the chip free surface can be close to that of the workpiece.
3. The heat stored in this layer combined with friction on the tool-chip interface cause high temperature at this interface. The tool side of the interface normally has higher temperatures than those on the chip side as the chip is moving while the tool rake face in stationary.

4. The highest temperature on the too chip interface is at the end of the plastic part of the tool chip contact that correspond to the maximum extend of the plastic deformation zone of the chip contact layer formed by the tool rake face.

1.3.5.5 Optimal Cutting Temperature Law

In deforming processes used in manufacturing, concern over the tool wear is often overshadowed by considerations of forces or material flow. Except for hot extrusion, die life is measured in hours and days, or in thousands of parts in metal-deforming operations [15]. In metal cutting, however, tool wear is a dominant concern because process conditions are chosen to give maximum productivity or economy, often resulting in tool life (the time to achieve the maximum allowable tool wear) in minutes. Central to the problem are: high contact temperatures at the tool-chip and tool-workpiece interfaces.

The nature of tool wear, unfortunately, is not yet clear enough in spite of numerous investigations carried out over the last 50 years. Although various theories have been introduced hitherto to explain the wear mechanism, the complicity of the processes in the cutting zone hampers formulation of a sound theory of cutting tool wear. Cutting tool wear is a result of complicated physical, chemical, and thermomechanical phenomena. Because different "simple" mechanisms of wear (adhesion, abrasion, diffusion, oxidation, etc.) act simultaneously with predominant influence of one or more of them in different situations, identification of the dominant mechanism is far from simple, and most interpretations are subject to controversy [15]. These interpretations are highly subjective and based on the evaluation of the cutting conditions, possible temperature and contact stress levels, relative velocities and many other process paramagnets and factors.

Standard tool-life testing and representation includes Taylor's tool life formula [32]

$$vT^n = C_T \tag{1.18}$$

where T is tool life in minutes, C_T is a constant into which all cutting conditions affecting tool life must be absorbed.

Although Taylor's tool life formula is still in wide use today and is in the very core of many studies on metal cutting including the level of National and International standards, one should remember that it was introduced in 1907 as a generalization of many-year experimental studies conducted in the 19th century using work and tool materials and experimental technique available at that time. Since then, each of these three components underwent dramatic charges. Unfortunately, the validity of the formula has never been verified for these new conditions. Nobody proved that far that it is still valid for any other cutting tool materials than carbon steels and high-speed steels, for higher cutting speeds than 10 m/min.

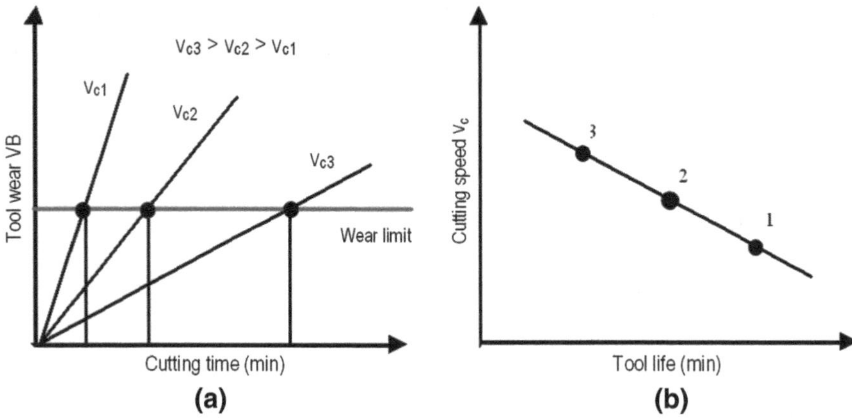

Fig. 1.15 Basic representation of the Taylor's formula: **a** Wear curves for several cutting speeds (1, 2 and 3), and **b** tool life curve

Simple analysis of the Taylor's tool life formula shows that it actually correlates the cutting temperature with tool life as the cutting speed uniquely determines the cutting temperature. Figure 1.15 shows common interpretations of the Taylor's tool life formula [33]. As seen, this formula states that the higher the cutting speed (temperature) the lower tool life which is in direct contradiction with well-known experimental studies and practice of metal cutting [8]. Leading tool manufactures clearly indicate the favorable range of the cutting speed (temperatures) for their tool materials. Deviation from the recommended speed (temperature) for a given tool material to either side lowers tool life. This, however, does not follow from the Taylor's tool life formula used in standards and practically all books (textbooks) on metal cutting.

To resolve this long-standing problem, the first metal-cutting law (Makarow's law after A.D. Makarow who fist pointed out the existence of the optimal cutting temperature) was formulated by Astakhov [1] in the following form:

For a given combination of the tool and work materials, there is the cutting temperature, referred to as the optimal cutting temperature θ_{opt}, at which the combination of minimum tool wear rate, minimum stabilized cutting force, and highest quality of the machined surface, is achieved. This temperature is invariant to the way it has been achieved (whether the workpiece was cooled, pre-heated, etc.).

The Makarow's law, established initially for longitudinal turning of various work materials, was then experimentally proven for various machining operations. Therefore, the optimum cutting temperature for a given combination "work material-tool material" should be established and used as the only criterion for the suitability of this particular tool material for this particular work material. This temperature is a physical property and thus does not depend on the intrinsic details of tool design and geometry as well as on the parameters of a particular test setup.

In these considerations, the cutting temperature is the temperature measured by the tool-work thermocouple technique. It has sense of the average or integral temperature of the tool-chip interface. The huge advantage of such a representation of the cutting temperature is that the tool-work thermocouple can be used in any metal cutting studies and even on the shop floor for practical machines with intelligent controllers that can easily measure and then control this temperature. The idea of the tool-work thermocouple and its calibration is well-known [1, 3, 22, 34]. Besides simplicity and feasibility of using on modern machine tools as the optimization criterion, a huge advantage of the optimal cutting temperature introduced this way is that it depends only on the work and tool material, and thus should be established only once to their given combination. Then, it is valid for any tool and tool design made of this tool material and for this work material regardless its heat treatment and particularities of the cutting operation.

Machining at optimal cutting temperature results not only in the minimum tool wear rate but also leads to obtaining the minimum cutting force and smallest roughness of the machined surface. Therefore, the proper "temperature gage" is developed for the first time in simulations of the metal cutting process. Achieving the optimal cutting temperature is the second criterion (the first—the minimum plastic deformation of the layer being removed) for optimization of the tribological process in metal cutting [1].

1.3.6 Issues to be Addressed

Although the discussed methodology allows to determine the mean shear and normal stresses at the tool-chip interface, and thus it provides some help in the selection of proper tool materials, coating and tool geometry parameters, there a few important issues that has to be pointed out here. They are: FEM modeling of the distribution of the contact stresses, tools with restricted contact length and high-speed machining effect on the contact stresses.

1.3.6.1 FEM Modeling of Contact Stress Distributions

Commercial FEM software packages (for example MSC.Marc, Deform2D, and Thirdwave AdvantEdge) are often used to model distribution of stress over the tool-chip interface. In such programs, the friction coefficient at the tool chip interface is an input variable so it should be known to the user/modeler/process or tool designer. As it is normally not known, the recommendation is rather simple—run the program at different friction coefficients and see what result is closer to what is in reality. Once 'acceptable' value of the friction coefficient is established then other output parameters of the machining process including contact stress distributions can be presumable obtained in such a modeling. In the author's opinion, however, there is a problem.

It is implied in Merchant's analyzes that the contact between the tool and the chip is a sliding contact where the coefficient of friction is constant [35]. In experimental studies, the following values for the friction coefficient were obtained: Zorev [8] obtained $\mu_{\mathrm{f}} = 0.6$–0.8, Kronenberg [23]—0.77–1.46, Armarego and Brown [36]—0.8–2.0, Finnie and Shaw [37]—0.88–1.85, Usui and Takeyama [38]—0.4–2.0, etc. An extensive analysis of the inadequacy of the concept of the friction coefficient in metal cutting was presented by Kronenberg (pp. 18–25 in [23]) who stated "I do not agree with the commonly accepted concept of coefficient of friction in metal cutting and I am using the term "apparent coefficient of friction" wherever feasible until this problem has been resolved." Unfortunately, it has never been resolved although more than 50 years passed since this Kronenberg's statement.

However, in the author's opinion, the problem is not in a particular value of the friction coefficient that varies even with a smallest change in the cutting process, and thus cannot be physically accounted for in FEM modeling having a value of the friction coefficient as an input parameter. The problem is that the concept of the friction coefficient sets the certain distribution of the contact stresses. It is explained as follows.

In most engineering and physical situations, friction effects at a tribological interface are described by a constant coefficient of Coulomb friction μ_{f},

$$\mu_f = \frac{F_f}{F_N} \tag{1.19}$$

Although it is well-established that contact between two bodies is limited to only a few microscopic high points (asperities), it is customary to calculate stresses by assuming that the forces are distributed over the total (apparent) contact area. Such an approximation, however, is not far from reality in machining where the actual and apparent contact areas are practically the same due to high contact pressures [8]. If it is so, the numerator and denominator of Eq. (1.19) can be divided over the tool-chip contact A_c and then recalling that the mean normal stress at the interface is $\sigma_c = F_N/A_c$ and the mean shear (frictional) stress at the interface is $\tau_c = F_f/A_c$, one can obtain

$$\mu_f = \frac{\tau_c}{\sigma_c} \tag{1.20}$$

Equation (1.20) reveals that if the friction coefficient at the tool-chip interface is constant, the ratio of the shear and normal stresses should be the same along the entire contact length. In other words, the distribution curves of the normal and shear stress MUST be equidistant over this length. Obviously, it is in direct contradiction with many experimental results obtained by various specialist and with practice of metal cutting [1]. Therefore, it is not clear how then meaningful results on the distribution of the shear and normal stresses over the tool-chip interface can be obtained using FEM software where the friction coefficient is an input variable.

1.3.6.2 Particularities of Stress Distribution on the Restricted Tool-Chip Contact Length

The majority of studies on metal cutting concerns with cutting tools having the full rake face, i.e. when the length of the rake face in the direction of chip flow is equal to or greater than that defined by Eq. (1.4). Although it is true for many drills and reamers, this is not normally the case for many single-point cutting tools with indexable inserts having chip breakers. In such tools, the length of the tool-chip contact is deliberately restricted to be smaller than the so-called "natural" contact length defined by Eq. (1.4).

According to Zhang [39], Klopstock in 1926 was the first to show that tool life and cutting forces could be favorably altered by restricting the tool-chip contact length. This was done using a composite rake face tool made of high speed steel. Klopstock found that the presents of the stable built-up edge results in better surface finish and greater tool life. Later on, it was found by multiple researchers that the use of tools with the restricted contact length may result in up to a 30 % reduction in the cutting force although the real reason for that is not clearly revealed. Limited-contact cutting has been studied by Takeyama and Usui [40], Chao and Trigger [41], Usui and Shaw [42], Hoshi and Usui [43] and Usui, Kikuchi, and Hoshi [44]. Detailed bibliography and analysis of the studies of this kind of tool were presented by Jawahir and Luttervelt [45] Luttervelt, Childs, Jawahir, Klocke, and Venuvinod [46], Zhang [39], Karpat and Ozel [47] and many others. The cutting mechanics for such tools was discussed by the author earlier in [1].

The most essential conclusions on the effects of the reduced contact length can be drawn from experimental results presented by Poletica [10] and Loladze [9], Zorev [8], Sadic and Lindstrom [48, 49]:

1. Tool life normally noticeably increases and the cutting force decreases when the tool-chip contact length is reduced from its natural length to the length of the plastic part of this contact defined by Eq. (1.2).
2. The rake angle of the restricted rake face is not an independent parameter. Rather, it affects the contact length through CCR ζ Eqs. (1.3) and (1.4).
3. Any further decrease of the tool-chip contact length beyond the length of plastic contact leads to rapid reduction of tool life.
4. The positive effect of the reduction of the tool-chip contact length becomes less profound for high cutting speeds.
5. When tool-chip contact length reduces, the maximum cutting temperature shift towards the cutting edge which in machining of difficult-to-machine materials leads to the plastic lowering of the cutting edge [1]. Figure 1.16

The foregoing analysis suggests that the maximum effect of the restricted tool-chip contact length is achieved when this length is equal to l_{c-p}, which, in turn, depends on the uncut chip thickness t_1 and CCR ζ. As t_1 is the direct function of the cutting feed and the tool cutting edge angle κ_r and CCR ζ is a function of the

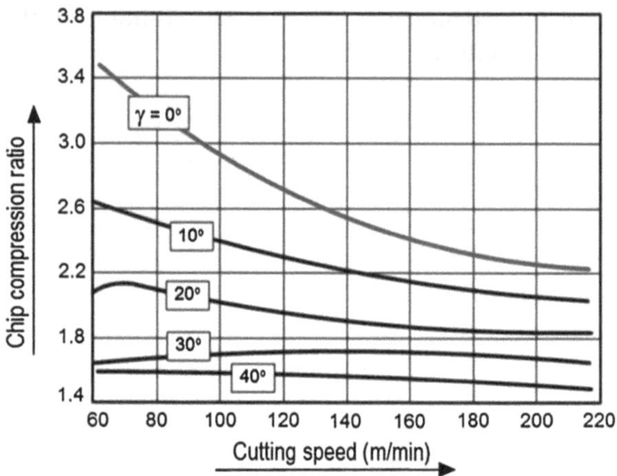

Fig. 1.16 Influence the rake angle on CCR for the range of the cutting speed used for carbide tools. Free cutting with $a_p = 6$ mm, $t_1 = 0.15$ mm. Work material: AISI steel 4130

tool and work materials properties as well as the cutting speed, feed and many other parameters of the machining system [1], this maximum effect can be achieved only for a specific applications when all these parameters are well known so that l_{c-p} can be determined with reasonable accuracy. Even small deviation from the optimal l_{c-p} may lead to significant change in tool performance. For example, Rodrigues and Coelho found [50] that the reduction of 0.25 mm in chamfer length and increase of 1° in chamfer angle (from SNMG PR to SNMG PF tools) caused a reduction in the specific cutting energy nearly 28.6 % and 13.7 % for conventional cutting speed and high-speed cutting respectively.

The vast majority of practical cutting tools including those with indexable inserts, however, are meant for wide ranges of the machining regime and various machining systems. Because these inserts have a fixed restricted contact length, the performance of these inserts may vary significantly depending upon a given application. This explains great scatter in the performance of indexable carbide inserts observed in practice. Understanding the concept of CCR and by measuring this important parameter in practical optimization of a cutting operation, any practitioner can select the proper insert for a given application.

1.3.6.3 Influence of the Cutting Speed

Many of 'classical' cutting test were carried out in the past using low cutting speeds [8]. The basic mechanics of metal cutting including distributions of the contact stresses over the tool-chip interface were obtained under this condition. For years, such approximation was satisfactory as cutting speeds used in practice have not result in qualitative change in the cutting mechanics. The time has changes and cutting speeds used today in modern industry are high that, in turn, resulted in qualitative change of the cutting mechanics. Unfortunately, the theory of metal cutting has not followed this trend so its basic principles of 'average values' are still constitute its very core.

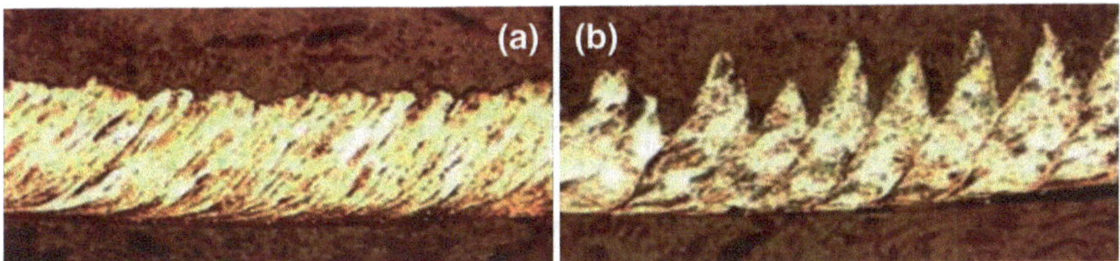

Fig. 1.17 Structures of the chip produced during machining of work a low-carbon annealed steel (0.04 %C), at **a** v = 150 m/min and **b** v = 1500 m/min (Courtesy of Prof. S. Ekinoviĉ)

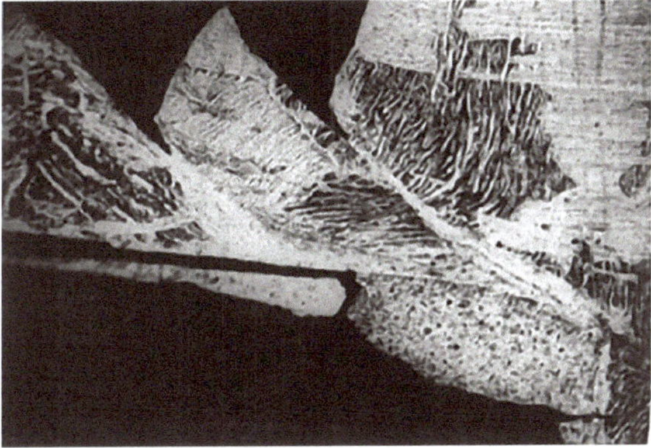

Fig. 1.18 Chip structure in machining of Ti-6Al-4 V alloy. Tool material carbide M10, cutting speed—150 m/min, depth of cut—1.5 mm, cutting feed—0.2 mm/rev

Figure 1.17 shows an example of qualitative change in chip structure at high cutting speed. Note that such a change occurred for one of the most ductile steel in the annealed (the softest) conditions. For other work materials, e.g. titanium alloys, such transition takes place even at lower cutting speed as shown in Fig. 1.18. As such, system considerations similar to that shown in Fig. 1.5 cannot be ignored. In other words, a significant variation of the cutting force takes place in each cycle of chip formation which results in corresponding changes in the distribution of contact stresses and temperatures. Naturally, the so-called 'static' distributions of contact stresses reported in the literature and their mean values are no more useful in optimization of cutting tool materials selection, design of cutting tool and optimization of the contact conditions at the tool chip interface. In the author's opinion, this is one of the major issues in the cutting tool tribology that has to be addresses as high-speed machining becomes an everyday machining process in modern industry.

1.4 Tribological Interfaces: Tool-Workpiece Interface

The results of the theoretical and experimental studies on the tribological conditions at the tool–workpiece interface can be summarized as follows [1]:

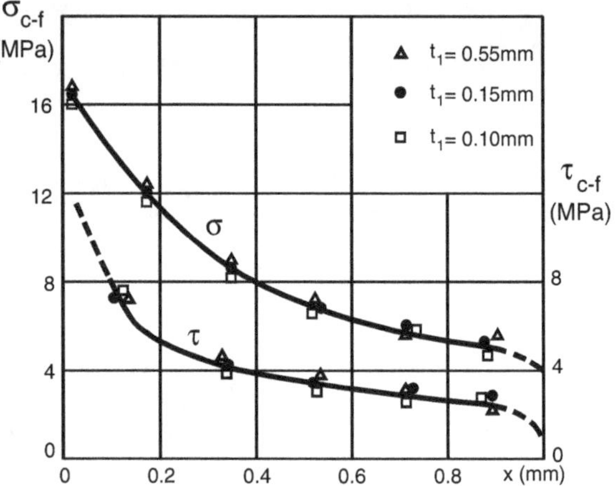

Fig. 1.19 Normal and shear stress distributions at the tool–workpiece interface in turning of lead with different uncut chip thicknesses, x is the distance from the cutting edge

1. The distribution of the normal and shear stresses have maxima in the region adjacent to the cutting edge. Then the level of stresses stabilizes over the contact length becoming zero at the end of the contact. The smaller the curvature of the workpiece surface, the higher the level of contact stresses at the end of the contact, where both stresses may have second maxima. Figure 1.19 shows typical distributions.
2. There is no or very small region of the plastic part of the interface.
3. Adhesion takes place in the region adjacent to the cutting edge while simple friction is the case on the rest of the interface.

The normal and shear stresses depend on the mechanical properties of the work material (primarily on its hardness), while other material characteristics, including material type, do not seem to have any noticeable effect. The stresses at the interface strongly correlate with the processes in the deformation zone and tool geometry through CCR. This fact should be briefly explained.

The surprisingly high contact stresses at the tool-workpiece interface cannot be explained even in principle by the common notion of spring back of the work material [1]. The proper explanation should include the bending moment due to normal force acting in the tool-chip interface as shown in Fig. 1.4. A model of the cutting wedge (the part of the cutting insert between the tool-chip and tool workpiece interfaces) deformation due to this moment is shown in Fig. 1.20. As shown, the normal force F_N causes the deformation of the cutting wedge that results in the 'diving' of the tool into the workpiece. Figure 1.21 shows experimental results for the penetration of the cutting wedge into the workpiece at the beginning of cutting as a function of the rake angle. As can be seen, this penetration is a strong function of the rake angle. This effect is known as "biting" of a sharp cutting tool with highly positive rake angle at the beginning of cutting. In the established cutting, the penetration of the cutting wedge into the workpiece occurs

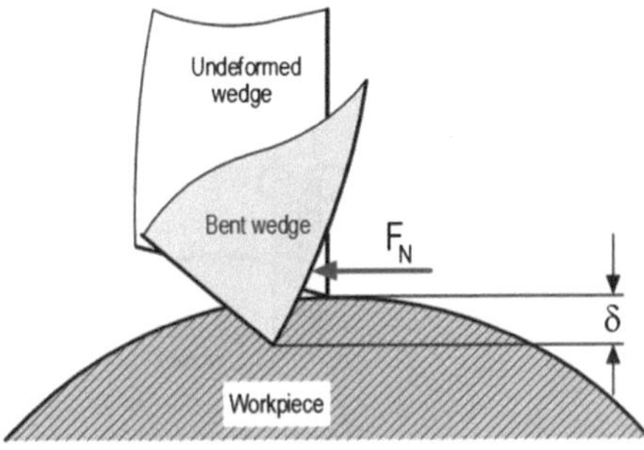

Fig. 1.20 Model of the cutting wedge deformation

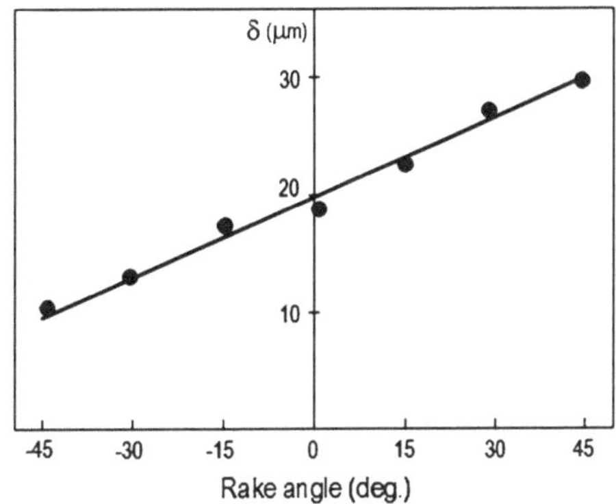

Fig. 1.21 Penetration of the cutting wedge into the workpiece at the beginning of cutting as a function of the rake angle. Carbide (P20) cutting insert CPMW having thickness of 3.5 mm. Work material, ANSI steel 1045, cutting feed $f = 0.25$ mm/rev

at the beginning of each chip formation cycle. It is in the range of 1–5 μ depending on the machining regime.

The obtained results explain the following known facts:

1. High contact stresses at the tool-workpiece interfaces that result in the wear of tool flanks.
2. Higher wear in machining of interrupted surfaces as these contact stresses are much higher at the beginning of cutting than at the beginning of each chip formation cycle.
3. The effectiveness of the tool with tangential insert location in the tool body. The principle of the so-called tangential tool design in its comparison with the conventional is shown in Fig. 1.22. Such a design was developed initially for tools working in heavy cutting conditions to prevent excessive tool wear. As can be seen in this figure, the bending of the insert in the tangential design is a way smaller than that in conventional because the thickness (the moment of inertia) of the insert is much greater.

Figure 1.23 shows examples of turning and milling tools with tangential cutting inserts by ISCAR Co. Similar designs gradually become available for other types of cutting tools, e.g. for drills.

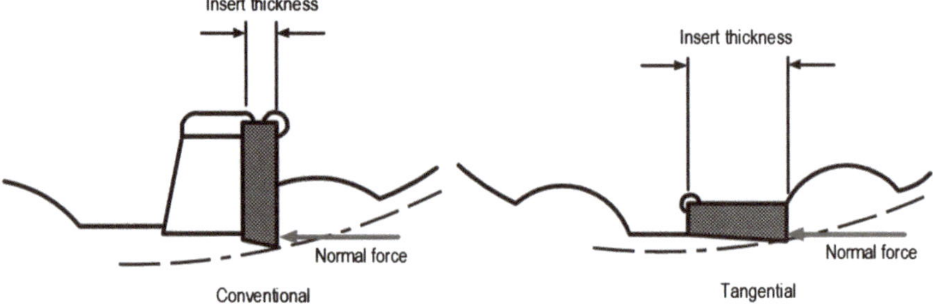

Fig. 1.22 Principles of conventional and tangential location of the cutting insert in the tool body

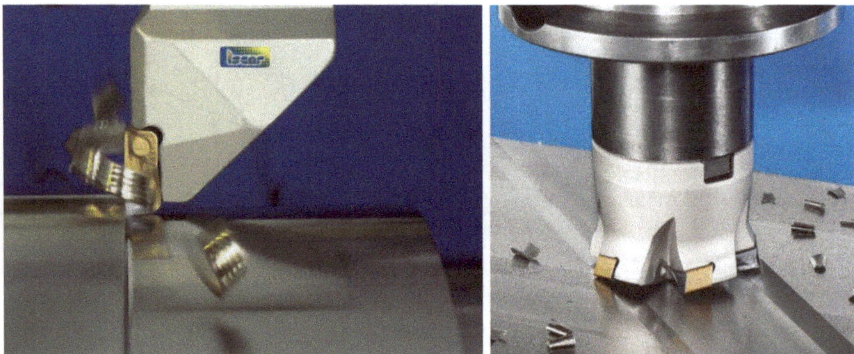

Fig. 1.23 Turning and milling tools with tangential cutting inserts (ISCAR Co.)

1.5 Tool Wear

Tool wear types and patterns are well described in the literature on metal cutting [1, 22, 51, 52]. The assessment and proper reporting of tool wear are standardized by International (e.g. ISO 3685:1993, ISO 8688-1: 1989) and National (e.g. ANSI/*ASME* B94.55 M-1985) standards. Tool wear is considered as a gradual process. Two basics zones of wear in cutting tools, namely flank wear and crater wear are normally conserved. Tool wear is most commonly measured using a toolmaker's microscope equipped with video imaging systems and having a resolution of less than 0.001 mm), stylus instrument similar to a profilometer, and with laser interferometer.

The general mechanisms that cause tool wear normally described in the literature are: (1) abrasion, (2) diffusion, (3) oxidation, (4) fatigue and (5) adhesion. These mechanisms are explained by many authors, for example, by Loladze [9], Shaw [22], Trent and Wright [16], Svets [34]. In addition to these 'standard' mechanism, Astakhov described and explained another mechanism called plastic lowering of the cutting edge that takes place in machining of difficult-to-machine materials [1, 51].

A new tribology-related issue related to cutting tool wear came over recently as some specialists realized that commonly listed tool wear types, as abrasion, adhesion,

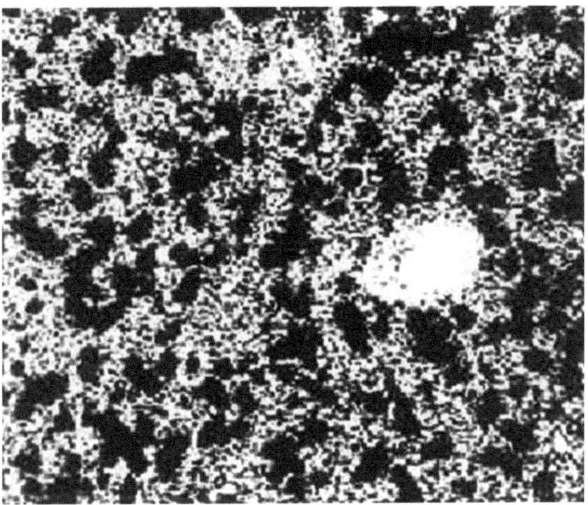

Fig. 1.24 Structure of cemented carbide with leached cobalt binder. In this condition, the skeleton of tungsten grains can break like glass

chemical, diffusion and other 'borrowed' from general machinery to describe tool wear, do not provide any help in identifying the root cause of tool failures. Although it has many facets as it manifests itself differently in various situations and thus it is called using variety of terms by practitioners, the proper name is cobalt leaching which can be caused by many factors. This is explained as follows.

Cobalt is by far most widely used binder metal or 'cement' in cemented tungsten carbides because it most effectively wets tungsten carbide grains during carbide sintering. For this reason cobalt is believed to be superior to other binder metals in terms of eliminating residual porosity and achieving high strength and toughness values in sintered products. However, a great disadvantage of cobalt as a binder metal is its leaching caused by various reasons. Except for some simple cases, such reasons are not yet well determined. The simplest yet rather common is corrosion due to chemical reaction of cobalt with corrosive agents. The corrosion process involves the dissolution of the cobalt binder at exposed surfaces leaving a loosely knit skeleton of tungsten carbide grains having little structural integrity. This mechanism is often referred as cobalt 'leaching' and is typically accompanied by flaking off of unsupported carbide grains in the affected surface areas. Figure 1.24 shows the appearance of cobalt leaching on a micrograph.

Although WC + Co grades have fairly good resistance to attack by acetone, ethanol, gasoline and other organic solvents as well as by ammonia, most bases, weak acids, and tap water, exposure to formic, hydrochloric, hydrofluoric, nitric, phosphoric, sulfuric, and other strong acids, however, can result in a relatively rapid deterioration of the binder phase. Corrosion rates are affected also by temperature, the concentration and electrical conductivity of the corrosive agent, and by other environmental factors as high contact pressure, relative speed of the solvents, etc.

According to the author's experience with carbide, tool failed due to corrosion-enhanced leaching and thus understanding of the issue, the carbide corrosion–

Table 1.2 Resistance to corrosion of cobalt and nickel binder cemented carbides

pH	Resistance to corrosion	
	WC + Co binder	WC + Ni binder
12	Very good	Very good
11		
10		
9	Good	
8		
7	Fair	
6	Poor	
5	Poor to no resistance	Good
4		
3		Fair
2		
1		Poor
0		

enhanced leaching can be explained as follows. There is a chemical process called *chelation*. Chelation is: (1) A chemical compound in which the central atom (usually a metal ion) is attached to neighboring atoms by at least two coordinate bonds in such a way as to form a closed chain, or (2) To cause (a metal ion) to react with another molecule to form a chelate. Dissolving would mean that the cobalt would break up into individual cobalt molecules in the water. Chelation means that it forms unique chemical compounds. This chelation causes the reddish or purplish coloration of a water-soluble MCF left around a failed tool. This coloration was observed by the author many times in investigating of failed tools.

Table 1.2 shows the corrosion resistant details of tungsten cemented carbide having cobalt nickel binder as function of the pH value of a water-soluble MCF which roughly explains the carbide corrosion–enhanced leaching. Although practically all books on MCFs (for example [53, 54] point out that pH 7–9 should be the target range in MCF maintenance, no proper explanation of this experience-gained knowledge is provided.

The author's experience shows that the described leaching occurs:

1. In actual part cutting when water-soluble MCFs are used. It is believed that triethanolamine (known as TEA) additive to many MCFs is primarily responsible for carbide leaching. When it happens, the wear pattern looks as a result of abrasion wear [55]. The matter gets worse when high-pressure MCF application to the rake face is used. Trying to improve one tribological condition, namely reduce the contact temperature, in such an application, other, namely the carbide extensive leaching, is enhanced.
2. In carbide manufacturing, particularly in carbide grinding with water-soluble MCFs.
3. In the stripping the old coating before the tool is re-ground and re-coated. It explains why carbide re-coated carbide tools show lower tool life than new tools.

Studies show (for example [56]) that the oleic acid triethanolamine ester solution could inhibit cobalt leaching in certain degree, boric acid ester solution could suppress it even better, and composite boric acid ester contain benzotriazole has the best inhibition effect. Its inhibition mechanism is that composite boric acid ester contain benzotriazole inhibit cobalt leaching of the cemented carbide tool by a layer of complete and compact protective membrane that generated on the surface of cemented carbide tool. Cobalt leaching of the cemented carbide tool is effectively inhibited by adding the composite boric acid ester contain benzotriazole in the water-based cutting fluid.

In the author's opinion and experience, the issue with cobalt leaching is a way wider that described above, It manifest itself in machining of reinforced plastics and other composite materials which quickly wear of the cutting tool having much greater hardness. The appearance of wear is as due abrasion, which is physically impossible. The analysis of the contact surface of worn tools showed severe cobalt leaching over the tool contact surface. It is caused presumable by strong bonding between the matrix material of the work composite material and cobalt in the cutting tool material. When sufficient volume of cobalt is leached, then collapse of skeleton of tungsten grains takes place so that wear pattern appearance resembles abrasion. Note that the author observed this mechanism not only for carbide but also for PCD tool material.

1.5.1 Improvements of Tribological Conditions of Cutting Tools

Many premature wear problems that create costly downtime facing many modern manufacturing plants today could be significantly reduced through some improvement of tribological conditions of the cutting tools used. To achieve profitability in the increasingly competitive global market modern manufacturing plants must operate with minimal downtime (i.e. avoid tool failures and premature tool pullouts due to quality problems of machined parts) while maintaining high efficiency of machining operations. These two objectives can be achieved simultaneously by analysis and optimizing the tribology of cutting tools.

The first step in selecting of particular improvement is to identify the mode of failure of the current tool. The failure analysis process aims to identify a particular or several modifications to the cutting tool to solve the problem. Any modification of tribological conditions in a cutting tool is considered as an improvement if it results in at least one of the following:

1. Improved tool life according to the criterion (or criteria when applicable) selected to evaluate this life for a given machining operation.
2. Improved productivity of a given operation by allowing a higher material removal rate, which directly depends on the cutting speed, cutting feed and depth of cut [6].

3. Increased efficiency of a given operation (cost per drilled hole, for example). In modern manufacturing, this improvement is often the most important as it is systemic improvement that affect the efficiency of the machining system as the whole. Unfortunately, many research papers, trade articles and promotional materials of tool companies do not consider this improvement, and thus methods of its achievement. Normally tool life is of prime concern, and thus the leading criterion for such an improvement.

All methods of improvement of tribological conditions can be broadly divided into two categories:

1. *Component means*. These methods include modification of component(s) of the cutting tool prior to the cutting process to improve its tribological conditions. Although there are a great number of these means, the following are common:

   1.1 Selecting a better type/grade of the tool material.
   1.2 Coating of the cutting tools with thin layer(s) of wear-resistant and/or friction-reducing (often referred to as tribological) materials [57].
   1.3 Designing cutting tools with application-specific macro- and micro-geometry [7], [58].
   1.4 Improving manufacturing quality of cutting tools.

Systemic means. These are all the components of the cutting system are actually engaged in cutting, i.e. when the cutting system is in existence. Among them, the following are directly related to the cutting tool tribology:

2.1. Application of the metal working fluid (MWF) through the cutting tool [59].
2.2. Improving tool holders.
2.3. Introduction of specially directed forced vibrations (often ultrasonic) into the cutting tool. Such an application may result in a decrease in the cutting force, better tool life, and, sometimes, in better integrity of the machined surface [60].

This section aims to discuss briefly the most vital aspects of various improvements of tribological conditions of cutting tools.

1.5.2 Grades of Tool Materials

The selection of cutting tool material type and its particular grade is an important factor to consider when planning a successful machining operation. A basic knowledge of each cutting tool material and its performance is therefore important so that the correct selection for each application can be made. Considerations include the type and properties of the work material to be machined, the component type and shape, machining conditions and the required quality for the considered operation. Note that the cost per machined part (the size of production lot, yearly program, existing machine/machining practice, etc.) should also be considered in the selection of the proper (technically and economically) tool material for a given machining application.

Starting considerations in the tool material selection are:

1. The kind/grade of tool material accounts for one-third of the tool successful performance.
2. The tool design, its manufacturing quality and proper implementation are two-thirds of the success.
3. Even the best salesman is not really an expert in your machining operation and system. The selection of the proper tool material including coating is your job for standard and especially for special cutting tools.
4. Test various tool materials/coating until you find what works for your application then keep testing for what works better.
5. There is always a hardness-toughness trade-off. Settle for as much toughness of the tool material as you absolutely have to have.
6. Select the allowable amount of tool wear (the criterion of tool life, for example the maximum with of the wear land on the flank face) as much as the quality of part (the component to be machined) and/or tool strength allow.
7. Although listed in multiple catalogs/books, often claimed by salesmen, maintained by wide-spread notions as equivalent, no two cemented carbide grades, superhard tool materials or anything else similar are exactly the same (besides the color and shape). The high production/technological culture, the better benchmarking in tool performance is established, the greater the difference.

1.5.2.1 Basic Properties of Tool Materials

Many types of tool materials, ranging from high-carbon steels to ceramics and diamonds, are used as cutting tool materials in today's metalworking industry. In modern drilling operations, three types of tool material are primarily used: (1) High speed steels (HSS), (2) Cemented carbides, (3) Diamond tool material including polycrystalline diamonds (PCD) and thick-film diamonds (TFCD) grades. In some special cases, for example in hard boring, polycrystalline boron nitride (PCBN) is also used. It is important to be aware that differences exist among tool materials, what these differences are in the correct application for each type of material [61].

The three general properties of a tool material are:

- *Hardness*: defined as the resistance to indenter penetration. It is directly correlates with the strength of the cutting tool material [62]. The ability to maintain high hardness at elevated temperatures is called hot hardness. Figure 1.25 shows the hardness of typical tool materials as a function of temperature.
- *Toughness*: defined as the ability of a material to absorb energy before fracture. The greater the fracture toughness of a tool material, the better it resists shock load, chipping and fracturing, vibration, misalignments, runouts and other

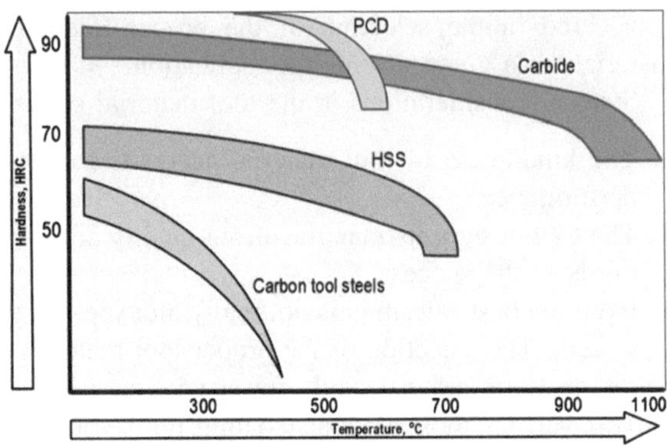

Fig. 1.25 Hardness of tool materials versus temperature

imperfections in the machining system. Figure 1.26 shows that, for tool materials, hardness and toughness change in opposite directions. A major trend in the development of tool materials is to increase their toughness while maintaining high hardness.

- *Wear resistance*: In general, wear resistance is defined as the attainment of acceptable tool life before tools need to be replaced. Although seemingly very simple, this characteristic is the least understood and, moreover, a subject to misinterpretation/misunderstanding.

Amongst these three characteristics, hardness is the simplest as many people have natural perception of hardness even not knowing its proper definition. Moreover, hardness testers of various kinds are widely available in many machine shops and test laboratories so that the measuring of this characteristic does not present any problems. The other two basic properties require some explanation as they explained considerably different in various sources of the professional literature. This should equip a specialist who tries to select a tool material with an ability to ask simple question about the test methods used to obtain these characteristics, and thus the relevance of the result in metal cutting.

The toughness of a hard tool material is determines using certain standard methods. For carbides, the short rod fracture toughness measurement is common, as described in the ASTM standard ASTM B771-11 "Standard Test Method for Short Rod Fracture Toughness of Cemented Carbides." The property K_{IcSR} determined by this test method is believed to characterize the resistance of a cemented carbide to fracture in a neutral environment in the presence of a sharp crack under severe tensile constraint, such that the state of stress near the crack front approaches tri-tensile plane strain, and the crack-tip plastic region is small compared with the crack size and specimen dimensions in the constraint direction. A K_{IcSR} value is believed to represent a lower limiting value of fracture toughness. This value may be used to estimate the relation between failure stress and defect size when the conditions of high constraint described above would be expected.

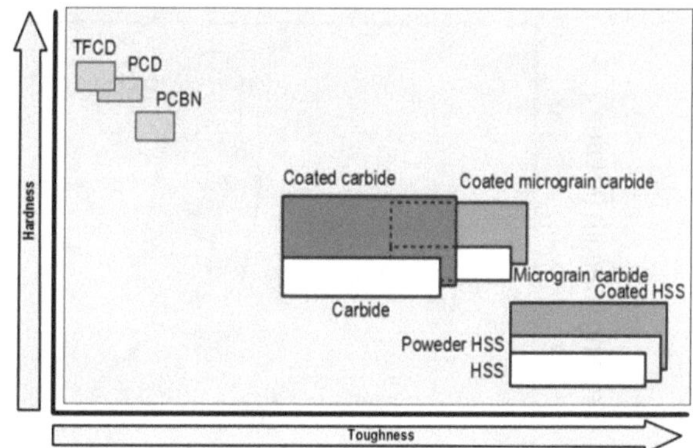

Fig. 1.26 Hardness and toughness of tool materials

Wear resistance is neither a defined characteristic nor a property of the tool material, but is instead a system response arising from the conditions at the sliding interface. The nature of tool wear, unfortunately, is not yet sufficiently clear despite numerous theoretical and experimental studies. Cutting tool wear is a result of complicated physical, chemical, and thermo-mechanical phenomena. Because various simple mechanisms of wear (adhesion, abrasion, diffusion, oxidation etc.) act simultaneously with a predominant influence of one or more of them in different situations, identification of the dominant mechanism is far from simple, and most interpretations are subject to controversy. Moreover, wear depends on the type of relative motion, normal stress, and sliding speed that bring a number of new variables in wear assessment. Because of these variations, different wear rates are commonly reported for the same combination of the tool and work materials in the literature.

Wear resistance should be somehow assessed to compare different tool materials, and thus helping a tool designer/end user in the selection of the proper tool material. The principle consideration in designing/using a wear test setup (rig, machine, methodology, and standard) is assuring that the test conditions, i.e. the sliding configuration conforms to the practical situation.

ASTM (respective subcommittees such as Committee G-2) standardized wear testing for specific applications, which are periodically updated. It developed a great numbers of standard wear tests starting with ASTM G40–10b "Standard Terminology Relating to Wear and Erosion". The Society for Tribology and Lubrication Engineers (STLE) has also documented a large number of frictional wear and lubrication tests. In any testing, wear should be expressed as loss of material during wear in terms of volume. The volume loss gives a truer picture than weight loss, particularly when comparing the wear resistance properties of materials with large differences in density. For example, a weight loss of 14.8 g in a sample of tungsten carbide + cobalt (density = 14800 kg/m^3) and a weight loss of 4.8 g in a similar sample of made of PCBN AMBORITE DBC50 (density = 4800 kg/m^3) both result in the same level of wear (1 cm^3) when expressed as a volume loss. The inverse of volume loss can be used as a comparable index of wear resistance.

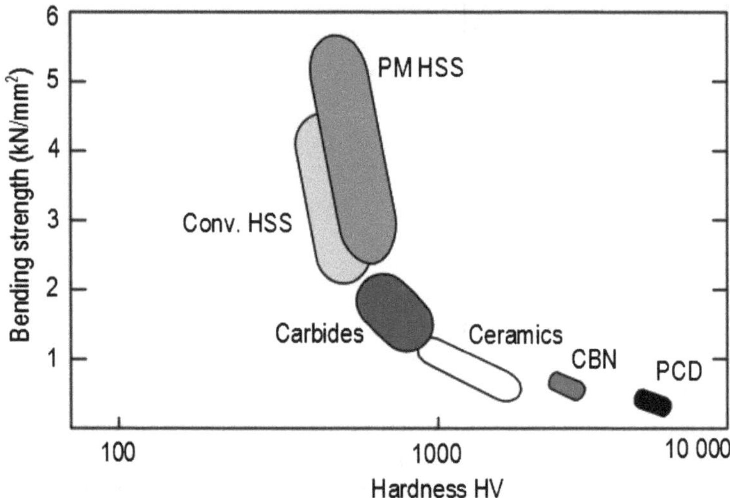

Fig. 1.27 Bending strength of tool materials

1.5.2.2 Selection of Application Specific Grade of HSS

High Speed Steel (HSS) today relates to a group of high-alloys, W-Mo-V-Co steels designed to cut other materials efficiently at high speeds. Once considered as an obsolete tool material which application range should reduce to hand and wood-cutting tools as new grades of other superior tool materials and new rigid machine are introduced, HSS survived as a high-performance tool material. Modern grades of HSS's combined with advanced coatings allow high cutting speeds that were considered 10 years suitable only for carbide tools.

HSS has the following advantages

1. Great bending strength which is significantly higher than any other cutting materials as can be seen in Fig. 1.27. It provides better resistance to cutting edge chipping, increased feed per tooth, and greater depth of cut.
2. Compare to other tool materials, a sharp cutting edge can be achieved even in conventional grinding. As a result, the followings are achieved: less work hardening of the work material that extremely important in machining of titanium alloys, austenitic stainless steels and nickel alloys, better surface quality, closer tolerances, lower cutting forces that particularly important in machining of thin-walled and non-rigid parts, lower cutting temperature, and thus smaller heat-affected layer in the machined surface.
3. Much greater tolerance for non-rigid machining system including old machines, non-rigid fixture, etc.
4. Much greater tolerance for specific work material/part conditions: nonhomogeneity of work material, cross holes, welding joints, inclined hole entrance, etc.

5. Great resistance to thermal shocks, and thus adaptation to practically all lubrication conditions.
6. Low cost per machined part.

In Fig. 1.27, PM HSS designated a group of powder metallurgy HSSs, which should be used to improve tribological conditions of HSS cutting tools. The fine structures of PM HSS that result from rapid solidification in the PM process offer premium characteristics for both the manufacturers of cutting tools and their users. The more uniform distribution and the finer size of carbides in PM steels are especially evident in comparisons with larger diameter bars of conventionally produced high speed steel, where carbide segregation is more of a problem. Thus, while the benefits pertain to cutters of all dimensions, they are more pronounced in larger tools.

There are four principal benefits of PM high speed steels for tool users. The primary benefit is the availability of higher alloy grades which cannot be manufactured by conventional steelmaking. These grades provide enhanced wear resistance and heat resistance for cutting tool applications. Second, the increased toughness of PM high speed steels not only provides greater resistance to breakage (particularly valuable in intermittent cut operations), but it also allows a tool to be hardened by 0.5–1.0 points higher on the Rockwell C scale without sacrificing toughness. Both longer tool life and higher cutting speeds can be realized. Third, PM HSS have improved grindability with no reduction in wear resistance of the tool. This means reduced grinding-wheel wear. Grinding can be done more quickly with less danger of damage to the cutter, and it leaves an edge that produces a smoother finish on the work piece.

Fourth, the greater consistency in heat treatment and uniformity of properties of PM HSS increases the degree of predictability for scheduling tool changes. This factor is particularly advantageous in multi-spindle machines, where a single cutter failure affects several spindles and usually requires changing all cutters (including some that may have a lot of life left) for the sake of prudence.

Figure 1.28 shows the various HSS grades, their chemical composition, properties, and availability. The large number of high-speed steels available today represents a number of choices based on performance and alloy modifications within the two major groups. The alloying modifications are based not only on tungsten and molybdenum contents, but also on carbon content, the total content of carbide-forming elements (including tungsten, molybdenum, chromium, and vanadium), and cobalt content.

Figure 1.28 lists not only the chemical composition of the commonly used HSS but also specifies the relative wear resistance, red hardness, toughness, and availability of 16 high speed steels. It should help a tool/process designer to select a suitable grade of HSS to improve tribological conditions, i.e. to address the identified failure mode of the cutting tool when this mode relates to a grade of HSS. For example, if it is identified that a threading tap made of M2 HSS and used for machining threads in high-silicon automotive aluminum alloy 380 suffers from excessive abrasion wear that is the prime mode of its failure then referring to Fig. 1.28, a significant improvement can be made if this tap is made of M3 TYPE 2 HSS.

DESIGNATION AISI (ISO)	NORMAL CHEMICAL COMPOSITION						ABRASION RESISTANCE	RED HARDNESS	TOUGHNESS	AVAILABILITY
	C	W	Mo	Cr	V	Co				
GROUP I - GENERAL PURPOSE										
M1 (HS 1-8-1)	0.80	1.50	8.00	4.00	1.00	—				VERY GOOD
M2 (HS 6-5-2)	0.85	6.00	5.00	4.00	2.00	—				EXCELLENT
M33	0.90	1.50	9.50	4.00	1.15	8.00				FAIR
M34	0.90	2.00	8.00	4.00	2.00	8.00				FAIR
M36	0.80	6.00	5.00	4.00	2.00	8.00				GOOD
T1 (HS 19-0-1)	0.75	18.00	—	4.00	1.00	—				GOOD
T2	0.80	18.00	—	4.00	2.00	—				POOR
T5	0.80	18.00	—	4.00	2.00	8.00				GOOD
GROUP II - SUPERIOR WEAR RESISTANCE										
M3 TYPE 1	1.05	6.00	5.00	4.00	2.40	—				GOOD
M3 TYPE 2	1.20	6.00	5.00	4.00	3.00	—				VERY GOOD
M7 (HS 2-8-2)	1.00	1.75	8.75	4.00	2.00	—				GOOD
GROUP III - HIGH RED HARDNESS HSS										
M41	1.10	6.76	3.75	4.25	2.00	6.00				FAIR
M42 (HS 2-9-1-8)	1.10	1.50	9.50	3.75	1.15	8.00				VERY GOOD
M43	1.25	1.75	8.75	3.75	2.00	8.25				VERY GOOD
GROUP IV - SUPER HSS										
M4	1.30	5.50	4.50	4.00	4.00	—				POOR
T15	1.50	12.00	—	4.00	5.00	5.00				EXCELLENT

Fig. 1.28 Basic HSS grades, their chemical composition, properties, and availability

1.5.2.3 Selection of Application Specific Grade of Cemented Carbides

The cemented carbides are a range of composite materials, which consist of hard carbide particles bonded together by a metallic binder. Cemented carbides consist of hard grains of the carbides of transition metals (Ti, V, Cr, Zr, Mo, Nb, Hf, Ta, and/or W) cemented or bound together by a softer metallic binder consisting of Co, Ni, and/or Fe (or alloys of these metals). Tungsten carbides (WC) known as straight grades are compounds of W and C. Because most of the commercially important cemented carbides contain mostly WC as the hard phase, the terms "cemented carbide" and "tungsten carbide" are often used interchangeably.

Cemented carbides belong to a class of hard, wear-resistant, refractory known as materials metal matrix composites in which hard tungsten carbide (WC) particles are bound together, or cemented, by a soft and ductile binder metal (commonly cobalt) known as the matrix. Although the term cemented carbide is widely used in the United States, these materials as known internationally as hard-metals.

The process of combining tungsten carbide with cobalt is referred to as sintering. On sintering, the process temperature is higher than the melting point of cobalt while it is much lower than that of WC. As a result, cobalt is embedding/cementing the WC grains and thereby creates the metal matrix composite with its distinct material properties. The naturally ductile cobalt metal serves to offset the characteristic brittle behavior of the tungsten carbide ceramic, thus raising its toughness and durability while lowering its hardness and thus abrasion wear resistance. Such parameters of tungsten carbide can be changed significantly by the cobalt content, grain size, addition of another (than WC) carbides, pre-sintering, sintering, and post-sintering processes, etc.

The proportion of carbide phase is normally 70–97 % of the total weight of the composite and its grain size averages between 0.2 and 20 μm. Tungsten carbide (WC), and cobalt (Co) form the basic cemented carbide structure and grades based on this concept are often referred to in simplified terms as straight grades. From this basic concept, many other types of cemented carbide have been developed. Thus, in addition to these simple WC–Co compositions, cemented carbide may contain varying proportions of titanium carbide (TiC), tantalum carbide (TaC) or niobium carbide (NbC) and others. Although they are called in the professional literature as, for example, titanium carbide (TiC), tungsten carbide (WC) is always predominate portion of such cutting tool materials.

Although cobalt as the matrix material can be alloys with or even completely replaced by other metals such as nickel (Ni), chromium (Cr), iron (Fe), molybdenum (Mo), or alloys of these elements, cobalt is the prime matrix material for cutting tool cemented carbides.

The first step in selecting the proper (for a given application) grade of cemented tungsten carbide is to identify the mode of failure on the current grade. To determine the actual mode of failure; e.g. abrasive wear, plastic lowering of the cutting edge, corrosion or mechanical failure (breakage), a failure analysis is carried out with some help by the carbide manufacturer. The failure analysis identifies the metallurgical and chemical properties of the current grade such as: binder material and content, grain size, hardness, density, transverse rupture strength (TRS), magnetic saturation and coercive force. A simple metallurgical study identifies the possible process control or quality issues, such as residual porosity (A, B or C type), binder lakes, clusters, eta phase, cross grade contamination, largest grain size and pits or voids. If needed, a test to determine if corrosion and other conditions such as mechanical fractures or green flaws are present is also conducted. After testing is complete, a comprehensive failure analysis report detailing causes of failure should identify the root cause of the problem. In the current context, tool carbide failure is considered as a condition when the tool does not perform as expected, i.e. to achieve the minimum cost per machine part. Among many possible root causes, the cutting tool tribilogy-originated failures becoming common in recent year as the overall quality of machining system became better as much tighter control of process variable is the case.

The starting point in the determination of the root cause of carbide tool material failure (in the introduced sense) is a series of micrographs of the tool material under consideration. Figure 1.29 shows a microstructure of cemented carbide at 1500x magnification. Such a magnification should be used for revealing the cutting tool tribilogy-originated failures.

Once the actual mode of failure and its root cause are identified, the selection of application specific carbide grade should not present any problem. When two key properties are controlled, manly the grain size and binder percentage, the tribological characteristics of the tungsten carbide grade can be predicted and consistent. The balance between the grain size of the tungsten carbide and binder material powders and the percentage of binder material must be achieved in order

Fig. 1.29 A microstructure of cemented carbide with a 12 % cobalt binder and 4 μ gran size at 1500x magnification

to get the desired combination of hardness, strength, toughness and shock resistance on a consistent basis. Figure 1.30 presents important information for the selection of the application-specific carbide composition.

Grain sizes are typically available in four categories; ultrafine with an average grain size of 0.5 μm or less, submicron with grain sizes of 0.8 μm, medium with average grain sizes of 1–2 μm and coarse at greater than 3 μm. Figure 1.31 shows the influence of grain size on carbide properties. Note that this picture is not strict as the different cobalt content is involved. If cosideren with the same cobal content, the picture will be considerably different.

The other key element that allows one to control the properties of the cemented tungsten carbide grade is the amount or percentage of binder material. The percentages of binder material will range from 3 % to approximately 25 % in cobalt binder grades and from 6 to 12 % when nickel is selected as the binder material. The general rule of thumb regarding binder percentage is the lower the percentage of binder material the harder and more wear resistant, but less the transverse rupture strength that directly correlates with impact resistance and toughness as shown in Fig. 1.32. As follows from this figure, the higher the percentage of binder material the higher the shock resistance or toughness, but less wear resistance.

When the grain size and percentage of binder material have been determined, the next consideration is the specific binder material. Cobalt is the most widely used binder material and, if abrasive wear is the problem to be solved, a cobalt binder material will be recommended. However, if severe adhesion (as to aluminum at high cutting speed) and/or corrosion are the issues, a nickel binder can be considered as an alternative. If problem persists then a nickel/chromium binder combination should be used. Note that sintered carbide with nickel and nickel/chromium binders are not on-shelf products so their cost is higher and lead time is longer. Therefore, they are to be used when no other means, as for example, coating, can solve the problem with leaching and/or adhesion.

GRADE	COBALT	TUNGSTEN CARBIDE	TITANIUM CARBIDE	TANTALUM CARBIDE	OTHER CARBIDES	HARDNESS HRA	ABRASION RESISTANCE	RED HARDNESS	TRANSVERSE RUPTURE STRENGH (TRS)
GROUP I - TUNSTEN CARBIDE - RESIST ABRASION (FLANK WEAR)									
C1	8.0	92.0				91.8			
C1	6.0	94.0				90.0			
C1	11.0	89.0				89.3			
C2	6.0	94.0				91.8			
C3	4.0	96.0				92.2			
C4	4.0	96.0				92.9			
GROUP II - TITANIUM CARBIDE BEARING GRADES - RESISTS CRATERING, SEIZING, GALLING									
C5	13.0	81.0	4.0	2.0		89.7			
C6	10.0	75.0	13.0	2.0		91.0			
C7	7.0	78.0	13.0	2.0		92.0			
C8	3.0	83.0	13.0		1.0VC	92.5			
GROUP III - TANTALUM CARBIDE (COLUMBIUM CARBIDE) AND TITANIOUM BEARING - RESISTS CRATERING PLASTIC LOVERING OF THE CUTTING EDGE UNDER HIGH TEMPERATURE AND NORMAL STRESS, SEIZING, GALLING									
C5	9.0	73.0	8.0	11.0	COL	91.0			
C6	7.5	74.5	8.0	10.0	COL	91.5			
C7	6.5	71.5	12.0	10.0	COL	92.0			
C8						92.5			
GROUP IV - TANTALUM CARBIDE BEARING GRADES - HIGH TEMPERATURE AND NORMAL STRESS RESISTANCE									
C5	12.0	55.0		28.0		86.0			
C5	11.0	21.0		18.0		89.0			
GROUP IV - HIGH VELOCITY - HIGH TITANIUM CARBIDE BEARING GRADES									
C8						92.1			

Fig. 1.30 Advanced carbide grades, their chemical composition and application properties

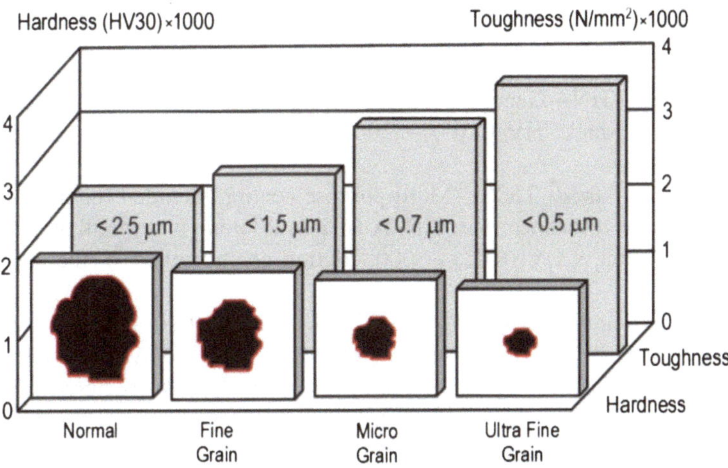

Fig. 1.31 Influence of grain size on carbide properties

1.5.3 Coating

HSSs and cemented carbides are excellent substrates for all coatings such as TiN, TiAlN, TiCN, solid lubricant coatings and multilayer coatings. Coatings considerably improve tool life and boost the performance of HSS tools in high productivity, high speed and high feed cutting or in dry machining, and machining of

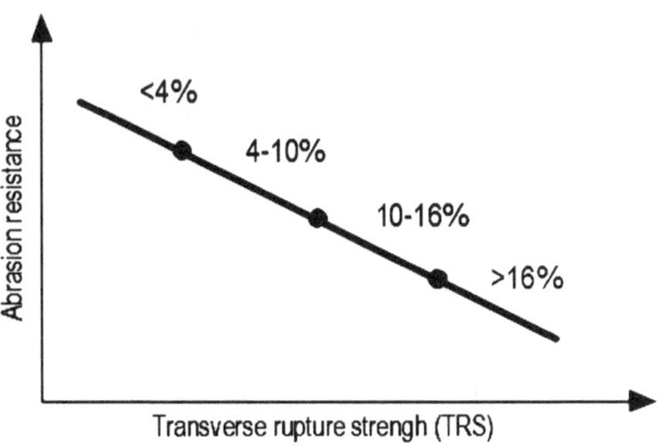

Fig. 1.32 Showing the influence of the binder amount

difficult-to-machine materials [63]. Coatings provide: (a) Increased surface hardness, for higher wear, (b) Resistance (abrasive and adhesive wear, flank or crater wear), (c) Reduced friction coefficients that easies chip sliding, reduces cutting forces, prevents adhesion on the contact surfaces, reduces heat generated due to chip sliding etc., (d) Reduced the portion of the thermal energy that flows into the tool, (e) Corrosion and oxidation resistance, (f) Crater wear resistance, (g) Improved surface quality of finished parts.

Common coatings applied in single or multi layers as shown in Fig. 1.33 [33]. They are:

Titanium Nitride TiN—General purpose coating for improved abrasion resistance. Color—gold, hardness HV (0.05)—2300, friction coefficient—0.3, thermal stability—600 °C.

Titanium Carbo-Nitride TiCN—Multi-purpose coating intended for steel machining. Higher wear resistance than TiN. Available in mono and multilayer. Color—grey-violet, hardness HV (0.05)—3000, friction coefficient—0.4, thermal stability—750 °C.

Titanium Aluminum (Carbo) Nitride TiAlN and TiAlCN—High performance coating for increased cutting parameters and higher tool life. Also suitable for dry machining. Reduces heating of the tool. Multilayered, nanostructured or alloyed versions offer even better performance. Color—black-violet, hardness HV (0.05)—3000-3500, friction coefficient—0.45, thermal stability—800–900 °C. Some companies, however, discontinued to the use of TiAlN as a standard coating. Instead much harder coating AlTiN is recommended for applications (AlTiN: 4500HV vs. TiAlN: 2600HV).

WC-C and MoS_2—Provides solid lubrication at the tool chip interface that significantly reduces heat due to friction. Has limited temperature resistance. Recommended for high-adhesive work materials as aluminum and copper alloys and also for non-metallic materials. Color—gray-black, hardness HV (0.05)—1000–3000, friction coefficient—0.1, thermal stability—300 °C.

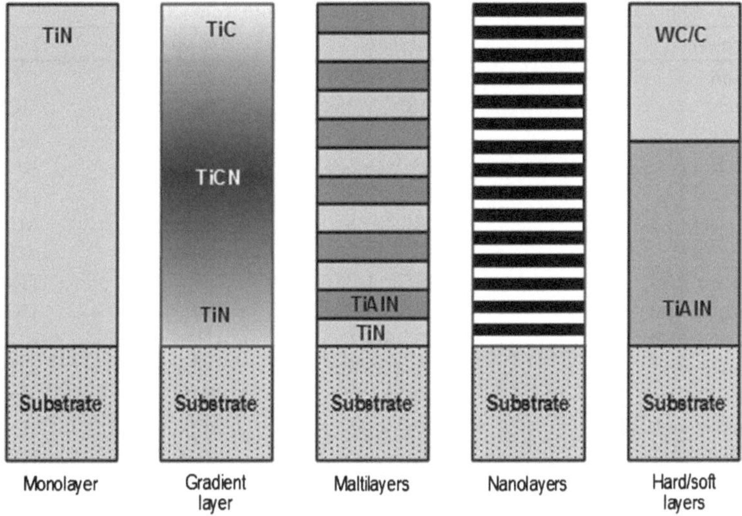

Fig. 1.33 Representation of layers of modern coatings on HSS tools

Chromium Nitride CrN—The anti-seizure properties of this coating makes it preferred in situations where built-up edge is common. HSS or carbide cutting will be seen with this almost invisible coating. Color—metallic.

Aluminum Titanium Nitride + *Silicon Nitride* nACo—hard coating for difficult-to-machine materials.

Diamond-Like Carbon (DLC). DLC has some of the valuable properties of diamond. When applied in pure form it is as hard as natural diamond. In pure form these diamond coatings offer extraordinary protection against abrasive wear and attack from atmospheric moisture and chemical vapors. Although smooth when seen with visible light, diamond like carbon actually has the form of a cobblestone street. In DLC the cobbles are not crystalline; they are amorphous because they are made from random alternations between cubic and hexagonal lattices. The cobbles have no long-range order and so they have no fracture planes along which to break. The result is a very strong material.

Table 1.3 shows the coatings recommended by Melin Tool Co for various work materials.

The first choice in thin-film coatings decades ago was TiN, the familiar yellow-gold coating still often used today. Some cutting tool end users still prefer old-fashioned TiN coating, unaware of the many choices now available. Dozens more have been developed in the meantime. Combined in multiple layers, coatings achieve a balance of properties not possible with a single-layer thick coating. As new thin-film ceramic coatings proliferate, the problem facing many engineers now may be one of too many choices, rather than too few.

Table 1.3 Application of coatings as recommended by Melin Tool Co

Work material	Hardness	1st choice	2nd choice
Aluminum	HB 160–240	TiCN	TiN
Alloy steel	HRC 23–38	TiN	TiCN
Alloy steel	HRC > 38	AlTiN	AlTiN
Carbon steel	HB 160–240	TiN	TiCN
Carbon steel	HRC 23–38	TiN	TiCN
Carbon steel	HRC > 38	AlTiN	AlTiN
Hardened steel	HB 130–240	nCAo	AlTiN
Low carbon steel	HRC 23–38	TiCN	TiN
Low carbon steel	HRC > 38	TiN	TiCN
Low carbon steel	HB 180–220	AlTiN	AlTiN
Gray cast iron	HB 220–320	AlTiN	AlTiN
Nodular cast iron	HB 180–220	nACo	AlTiN
Austenetic stainless steel	HRC < 35	TiCN	TiN
Martinsitic stainless steel	HRC ≥ 35	AlTiN	TiCN
Martinsitic stainless steel	HRC < 35	nACo	AlTiN
Ni alloys	HRC ≥ 35	nACo	AlTiN
PH stainless steel		AlTiN	TiCN
PH Stainless Steel		nACo	AlTiN
Ni, Co, Fe Based superalloys		AlTiN	AlTiN
High si aluminum		nACo	AlTiN
High si aluminum		nACo	AlTiN

Conventional coatings having rather rough surface are not inherently lubricious. As a result, applications of such coating solved problem with tool protection but gave rise to another problem of higher heat generation on the tool-chip and tool-workpiece interfaces. New coatings have been developed to address this problem. Multilayered coatings combine lubricity and resistance to wear. This combination reduces friction at the tool-chip and tool-workpiece interfaces.

A noticeable trend in tool coatings is that they are becoming application specific. For example, Oerlikon Balzers (Amherst, NY), a supplier of coatings and surface technologies, now offers 24 different coatings. The Oerlikon Balzers developed application-specific coatings such as the Balinit Aldura for milling hardened tool steels with hardness exceeding R_C 50 and Balinit Helica, an AlCr-based coating designed specifically for twist drills. Balinit Aldura uses an aluminum-chromium-nitride (AlCrN) based functional layer deposited onto a TiAlN support layer, all on a carbide substrate. The TiAlN ensures good adhesion and mechanical strength, while the AlCrN layer features excellent hot hardness and oxidation resistance (up to 1100 °C) and insulates the tool from the heat of cutting.

Although the listed coatings and the know coating application techniques seemingly work for indexable cutting insert which are not subjected to regrinds, a combined cost, logistic and tool life problem arises for cutting tool, as for example drilling tools, that have to be re-ground. The only proper way to carry out re-grinding

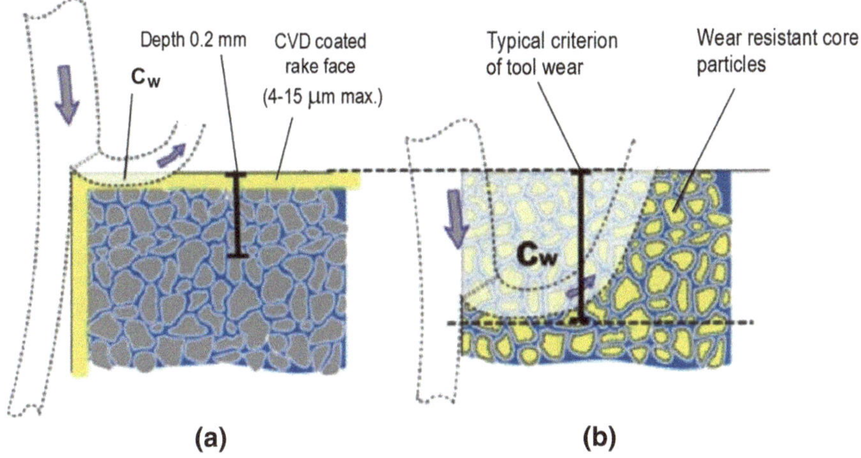

Fig. 1.34 Representation of: **a** Conventional CVD coating, and **b** TCHP

includes stripping the old coating, re-grinding the tool, re-coat the tool. It adds lead time and cost associated with re-grinding. Moreover, reground tools often have lower tool life compare to new tool due to cobalt leaching on coating stripping.

To solve this problem, a considerable different way of coating, namely the coating of carbide powder rather than coating the finish carbide product, can be used. It is known for long time but only recently became feasible for practical applications, can be used. The complete literature review on the development history of this method can be found in [64].

One realization of this concept is Tough Coated Hard Powders (TCHPs) which are a new family of patented, high performance metallurgical powders that incorporate unprecedented combinations of property extremes. They represent a class of engineered microstructure P/M based hardmetals having combinations of critical properties that provide improvements in performance and productivity. These engineered property combinations include toughness, abrasive and chemical wear resistance, low coefficient of friction, and light weight and so on at levels not previously seen. TCHP powders can be fabricated into a multitude of industrial metal-cutting inserts to leverage their key attributes to achieve manufacturing productivity improvements. These TCHP powders are created by incorporating hard particles in a tough matrix using proprietary manufacturing technologies. Engineered nanostructures are designed by encapsulating extremely hard "core" particles with a tough outer layer(s), for example tungsten carbide and cobalt, which in the consolidation process becomes a contiguous matrix.

Simplified comparison of performances of the conventional Chemical Vapor Deposition (CVD) coating and TCHP are shown in Fig. 1.34. As can be seen in Fig. 1.34a, when the wear of the rake face reached 0.2 mm which is the most common criterion of tool wear in finishing operations, no coating even in traces

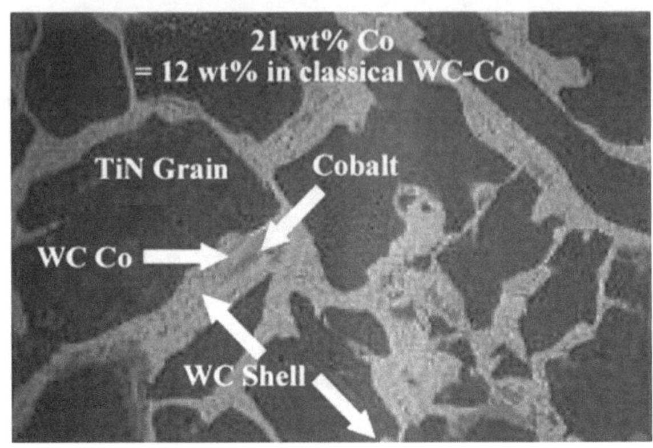

Fig. 1.35 Theoretical and actual macrostructures of a TCHP grade insert

can be found on the rake face to protect the tool against wear. Increasing the thickness of conventional coating is limited by delamination and cracking from different thermal expansion rates over large areas, bending, and surface loads severely limit coating thickness. In contrast, TCHP sintered microstructure is a cellular pseudoalloy of a contiguous tough tungsten carbide and cobalt mechanical support and binder phase containing chemically unadulterated wear-resistant core particles. These particles (such as TiN, TiC, TiB2, ZrN, Al2O3, diamond, cBN, AlMgB14, or B4C) are dispersed evenly throughout the tough tool. When using multiple core particle materials to enhance different properties, each material and its unique properties are available simultaneously at the working surfaces and cutting edges of the tool throughout the entire substrate. This design multiplies the volume of wear resistant material useable many times than in any possible coating, providing a continuously renewed (self-healing) wear surface. Figure 1.34b shows that the depth of wear used as the criterion of tool wear in roughing operation. The TCHP work all the tome this criterion is achieved so tool life can be extended many times.

TCHP powders and consolidated carbide blanks are manufactured and sold by Allomet Corporation (North Huntingdon, PA) as EternAloy®. Representative "core" particles include those traditionally used for extreme wear resistance (e.g., diamond, cBN, Ti(C,N), TiN, Al2O3). Example microstructures of a cutting insert of TCHP grade are shown in Fig. 1.35.

TCHP "composaloys" allow new combined levels of fracture toughness and hardness; resistance to abrasion, friction, wear, and corrosion; thermal conductivity; and impact resistance throughout the entire tool material. This innovation may extend uncoated tool life 10–30 times and coated tool life approximately four to seven times. In addition, the tool material can be reground and recycled an additional one to five times before disposal. Naturally, tools made with TCHP "composaloys" do not require an external coating because hard and tough phases are already dispersed throughout the tool, resulting in a continuously renewed wear surface within a tough substrate.

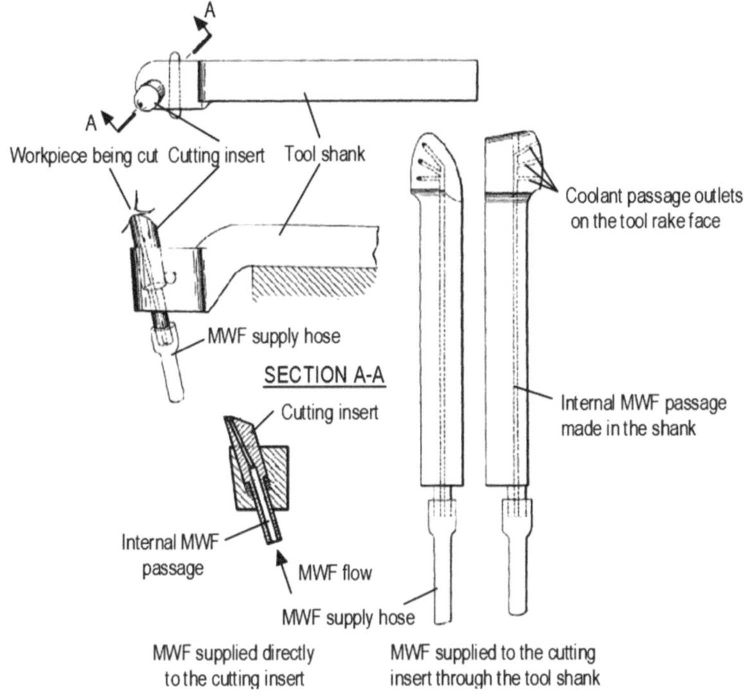

Fig. 1.36 Single-point cutting tool with MWF-through application (US Patent 160,161 (1875)) directly to the cutting insert and to the tool rake face through the tool shank

1.5.4 Application of the Metal Working Fluid Through the Cutting Tool

The application of MWFs is normally considered by the tribology of metal cutting as such an application a component of metal cutting system. Moreover, as discussed by the author earlier [65], MWF normally affects the properties of the work material, and thus reduces the energy of plastic deformation of the layer being removed. However, when MWF is applied at high pressure directly to the cutting tool tribological interfaces and through the cutting tool, the MWF actions may directly affect the tribological processes on these interfaces. Therefore, the application of MWF is considered in this chapter in this sense, i.e. as applied at high pressure aiming to alter the tribological process at the tool-chip and tool-workpiece interfaces.

Although the conventional perception is the through-tools MWFs technology is a relatively recent technology, attempts to use this technology to improve tool life and quality of the machined surface were made since the end of the 19th century. As an example, Fig. 1.36 shows a single point tool with an internal MWF supply into the tool rake face. As can be seen, it includes all the components of high-pressure through-tool MWF supply systems used today in the most advanced machining centers. In other words, this design was reinvented many years later when high-pressure pumps became widely available.

With a through-MWF system, MWF is usually pumped through the tool and then through the outlet MWF orifice (nozzle) to the machining zone. There are three principle objectives in using this technology: (1) increase tool life, (2) facilitate chip breakage and chip removal, and (3) assure dimensional stability of the machined surface (particularly important for finish reamers, for example, in the automotive industry). Depending upon which one of these three objectives is more important for a particular application, a specific method of MWF-through supply is selected.

The beginning of the systemic studies of this technology is attributed to work of Pigott and Colwell [66] who conducted an experimental study in turning of SAE 2150 work material with a high speed steel tool where a jet of MWF having pressure up to 2.76 MPa was directed into the tool-workpiece interface. They showed that in the range of 'normal' cutting speed, tool life can be extended up to eight times. According to the conclusions made, the tougher the work material, the higher the cutting speed, the greater improvement in tool life (up to 30 times) can be achieved. Moreover, the broken rectangular chips were obtained and the surface finish of machined parts was improved. Probably the most important conclusion made in this study is that the positive results are achieved only for a proper combination of the tool materials and design, work material, machining regime, MWF jet parameters and other system characteristics of the machining system. Unfortunately, the subsequent researches [67–77] (with some rare exceptions, where attempts were made to understand the essence of the technology [78, 79]) did not pay much attention to this important conclusions.

For years, there are a number of articles and papers are written to promote high pressure MWF application. In many of these publications, stunning results are presented: tool life increases up to 20 times, cutting speed can be increased up to three times, chip breaking is great even in machining of most difficult-to-machine materials known to produce difficult to break chips in the form of long spirals or unbreakable strings. A logical question is about why this seemingly attractive technology has not yet found a wide application in metalworking industry despite its obvious advantages. In the past, many machine tools and cutting tools did not support this technology. Nowadays, modern machine tools have complete enclosure of the working space that combined with intelligent controllers assure no danger even if an extremely high MWF pressure is used. Wide variety of modern cutting tools is available with high-pressure nozzles. Figure 1.37 shows examples of cutting tools for high-pressure MWF application by Sanvik Coromant Co and Fig. 1.38 shows examples of those by ISCAR Co. Some other companies as, for example, ChipBlaster (PA, USA) offered a complete set of high-pressure supply products to retrofit a machine. High-pressure MWF supply systems became almost standard units that can be bought, and thus a machine can be retrofitted for high-pressure MWF supply at very reasonable costs. What seems to be the problem with wide implementation of this technology?

Fig. 1.37 Boring bar and milling tool with high-pressure MWF supply by Sandvik Coromant Co

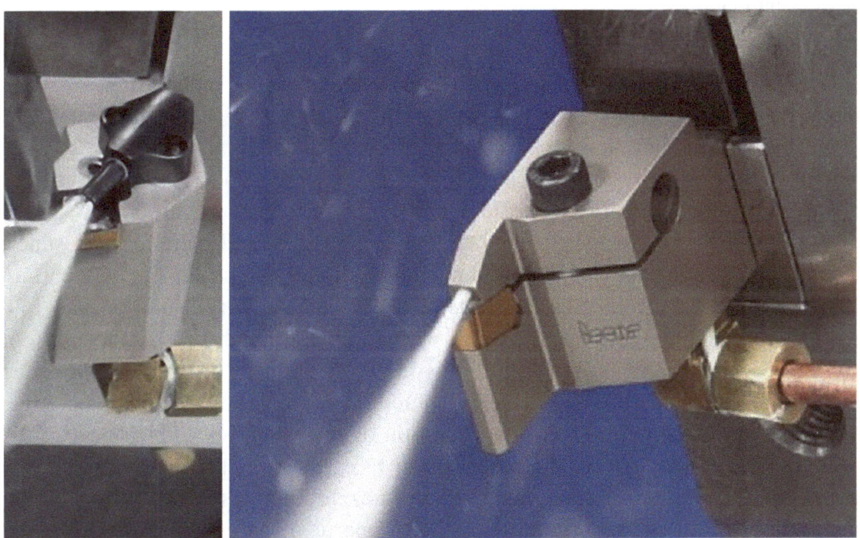

Fig. 1.38 Examples of cutting tools with high-pressure MWF supply by ISCAR Co

The answer to this question is simple: high-pressure MWF is highly application specific, and this its results vary significantly from one to the next seemingly "similar" machining operation. Reasonable explanations for this 'similarity' have not yet been provided. The known research papers offer the results obtained for considerably different machining conditions, work materials, MWF pressures and nozzle designs so that it is next to impossible to compare their results, and thus come out with rational recommendation for the application particularities of this technology.

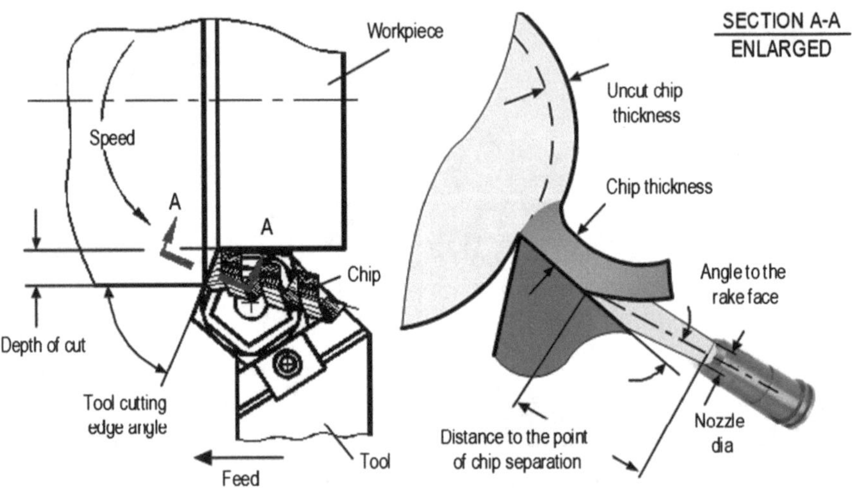

Fig. 1.39 Visualization of the variables involved in a high-pressure MWF application in a single-point tool

There are two prime objectives of implementing this technology, namely, improving chip breaking and increasing tool life. These two should be considered independently as they may not be achieved simultaneously as governed by considerably different physical and mechanical phenomena. The variables involved, however, are the same. Figure 1.39 visualizes some important variables.

For a given work material (its chemical composition and metallurgical state), the effect of high-pressure MWF application directly depends on: (1) cutting speed, (2) the uncut chip thickness defined by the feed per revolution and tool cutting edge angle, (3) the chip width defined by the depth of cut and the tool cutting edge angle, (4) the chip flow direction defined by the cutting edge inclination angle, (5) the tool material including coating, (6) the tool rake angle and rake face configuration, (6) the direction of the MWF jet with respect to the tool rake face (the angle to the rake face in Fig. 1.33), (7) the distance from the nozzle to the chip separation point, (8) the nozzle diameter and MWF pressure that define the MWF flow rate, (9) MWF properties. Moreover, there are many complex interrelations among the listed parameters. For example, the location of the point of chip separation is determined by the uncut chip thickness (which in turn depends on the feed per revolution and the tool cutting edge angle), the chip compression ratio (which is turn is a complex function of the machining regime, tool rake angle and its rake face configuration as well as the properties of the tool and work materials), and the cutting edge inclination angle. As a result, it is next to impossible to optimize all of these parameters particularly for a tool that performs variety of machining operations. A seemingly small change in the listed parameter may affect the result of high-pressure MWF application dramatically. This explains a great scatter in the results reported in the literature.

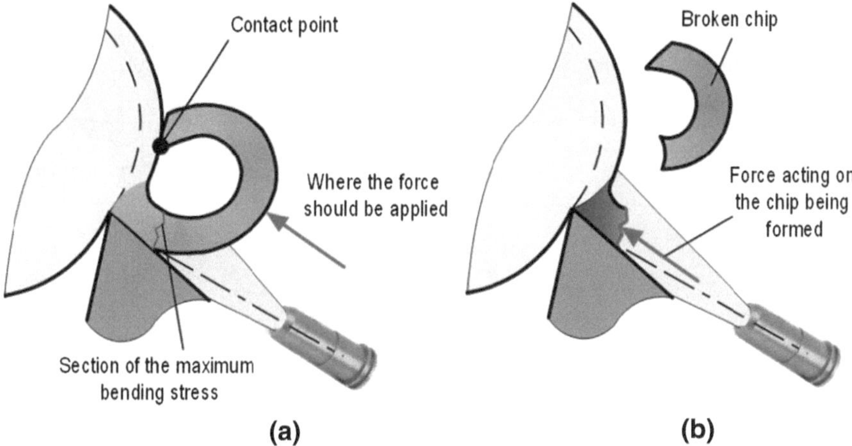

Fig. 1.40 Simplified model of chip breaking in high-pressure MWF application: **a** initial stage, and **b** formation of a new chip element

Out of three prime objectives of high-pressure MWF application, chipbreaking is the simplest although its achieving requires some considerations in terms of the nozzle location and MWF flow rate (nozzle diameter and MWF pressure) provided that the properties of the work material and the uncut chip thickness (or at least its range) are known. When properly applied, this technique assures reliable chip removal.

A good start for practical considerations is a combination of the following items:

- Chip forms classification: Standard ISO 3685 1993
- Chip structure classification based on the mechanism of its formation that essential to the chip breakage [12, 80].
- Chip flow direction as determined by the cutting tool geometry [7, 81].
- The fundamental of chip control with various chip breakers [45].

Figure 1.40 shows a simplified model. As shown in this figure, the most common way to apply high-pressure MWF is to aim the center of the MWF jet at the interface between the chip and the rake face. Although the usual explanation is that the MWF applied this way facilitate MWF penetration into the tool-chip interface, it is not so as the contact pressure at this interface is a way too high to achieve this ambitious goal [1]. Rather, high pressure MWF directed at the tool chip interface creates a 'wedge' as shown in Fig. 1.41 that facilities chip bending. When the stress at the section of the maximum bending stress (Fig. 1.40a) reaches a certain work-material-specific limit, the chip breaks over this section. Often, it happens not due to direct action of MWF but when the chip being formed touches a surface that moves at different velocity than that of this chip. It is shown Fig. 1.40a that the chip being formed touches the so-called transient surface of the workpiece. In practice however, it can also be the surface to be machined or the

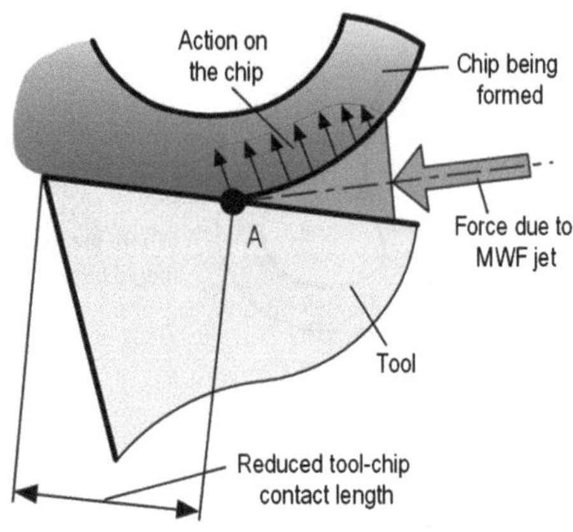

Fig. 1.41 Chip bending under high pressure MWF application into the tool-chip interface

machined surface of the workpiece, tool shank, fixture, etc. depending on the tool geometry and machining regime.

To achieve the highest mechanical action of the MWF high-pressure jet on the chip, the center of the jet, however, should be applied where is creates the maximum bending stress as shown in Fig. 1.40a. Unfortunately, such an application works only for the first chip element. Then, the second chip element would have a hard time to pass through high-pressure coolant jet that can ruin the whole idea of chipbreaking. Therefore, although mechanically it is not efficient to create the bending stress in the section of the maximum bending stress, stable chip breaking is achieved when the MWF jet is applied as shown in Fig. 1.40a. Even with such an application, the pressure of the MWF jest acts to resist chip formation (Fig. 1.40b) that creates an additional unnecessary force. Therefore, the MWF jet should be directed toward the rake face with the center below the point of chip separation. As such, however, there two prime concerns: (1) a great deal of MWF energy is lost on the MWF jet interaction with the tool rake face, and (2) the point of chip separation is not exactly known being dependant on many process parameters. Moreover, this point shifts towards the cutting edge due to the mechanical action of the MWF jet and this shift also depends on the process parameters and the jet flow rate. Therefore, conservative estimation of the separation point should be made. Such an estimation, however, is difficult when the inserts with complex chipbreakers made on the rake face is used.

Nowadays, a great variety of new inserts with complex chip breakers are available that can handle chipbreaking for various work materials and great range of machining regimes. It makes high-pressure MWF applications a less attractive option for such a purpose. It should be noted, however, that when properly designed and used, this technique assures reliable chipbreaking for much wider range of machining regimes and machining operations compare with even most advances design of chipbreakers made on cutting inserts. Therefore, for a given operation(s) (station in a manufacturing cell), simple feasibility analysis should be carried out to compare available chipbreaking options. In performing such an analysis, one should keep in mind that chipbreaking due to high pressure MWF application does not increase the cutting force, normally improves tool life and surface integrity [82] of machined surfaces compare to chipbreakers made on cutting inserts.

The mechanism due to which tool life increases in high-pressure MWF application is not obvious. As discussed by the author earlier [65], high-pressure MWF application may bring the cutting temperature closer to the optimal cutting temperature, and thus tool life increases. This is the cooling action of this technique. In the author's opinion, however, another mechanical action of high-pressure MWF application which affects tool life should be considered. This action is in changing the tool-chip contact length.

As discussed in Sect. 1.3.2, when tool-chip contact length reduces, the maximum cutting temperature shift towards the cutting edge which in machining of difficult-to-machine materials leads to the plastic lowering of the cutting edge [1]. Moreover, as the cutting power needed to machine a given material does not significantly change, the normal contact stress at the tool-chip interface may increase dramatically that may significantly reduce tool life when using the tool material which susceptible to chipping under high contact stresses. For example, Ezugwu reported [83] that tool life in machining of Inconel 718 with SiC whisker reinforced alumina ceramic tool reduced by 50–60 % compare to conventional flood MWF application due to micro chipping of the tool material while a great increase in tool life under the same condition was achieved when a coated carbide was used as the tool material.

The foregoing analysis suggests that the maximum effect of the restricted tool-chip contact length is achieved when this length is equal to $l_{c\text{-}p}$, which, in turn, depends on the uncut chip thickness t_1 and the chip compression ratio ζ. As t_1 is the direct function of the cutting feed and the tool cutting edge angle κ_r [7] and CCR ζ is a function of the tool and work materials properties as well as the cutting speed, feed and many other parameters of the machining system, this maximum effect can be achieved only for a specific applications when all these parameters are well known so that $l_{c\text{-}p}$ can be determined with reasonable accuracy. Even small deviation from the optimal $l_{c\text{-}p}$ may lead to significant change in tool performance. It explains a wide scatter in the results on the implementation of high-pressure MWF technique. Moreover, nozzle design and its location, MWF brand and pressure also add to this scatter. Therefore, to achieve real and sustainable benefits of this technique, systemic studies on the subject are needed followed by detailed implementation instructions.

References

1. Astakhov VP (2006) Tribology of metal cutting. Elsevier, London
2. King RI, Hahn RS (1986) Handbook of modern machining technology. Chapman and Hall, New York
3. Astakhov VP (1998) Metal cutting mechanics. CRC Press, Boca Raton

4. Astakhov VP, Xiao X (2008) A methodology for practical cutting force evaluation based on the energy spent in the cutting system. Mach Sci Technol 12:325–347
5. Astakhov VP, Shvets S (2004) The assessment of plastic deformation in metal cutting. J Mater Process Technol 146:193–202
6. Astakhov VP (2011) Turning. In: Davim JP (ed) Modern machining technology. Woodhead Publsihing, Cambrige
7. Astakhov VP (2010) Geometry of single-point turning tools and drills. Fundamentals and practical applications. Springer, London
8. Zorev NN (1966) Metal cutting mechanics. Pergamon Press, Oxford
9. Loladze TN (1958) Strength and wear of cutting tools (in Russian). Mashgiz, Moscow
10. Poletica MF (1969) Contact loads on tool interfaces (in Russian). Mashinostroenie, Moskow
11. Abuladze NG (1962) The tool-chip interface: determination of the contact length and properties (in Russian). In: Machinability of heat resistant and titanium alloys. Kyibashev Regional Publishing House, Kyibashev (Russia)
12. Astakhov VP, Svets SV, Osman MOM (1997) Chip structure classification based on mechanics of its formation. J Mater Process Technol 71:247–257
13. Astakhov VP (1999) A treatise on material characterization in the metal cutting process. Part 2: cutting as the fracture of workpiece material. J Mater Process Technol 96:34–41
14. Spaans C (1972) Treatise on the streamlines and the stress, strain, and strain rate distribution, and on stability in the primary shear zone in metal cutting. ASME J Eng Ind 94:690–696
15. Schey JA (1983) Tribology in metalworking. American Society for Metals, Metals Park
16. Trent EM, Wright PK (2000) Metal cutting, 4th edn. Dutterworth-Heinemann, Boston
17. Childs THC, Maekawa K, Obikawa T, Yamane Y (2000) Metal machining. Theory and application. Arnold, London
18. Astakhov VP (2005) On the inadequacy of the single-shear plane model of chip formation. Int J Mech Sci 47:1649–1672
19. Schmidt AO, Gilbert WW, Boston OW (1945) A thermal balance method and mechanical investigation of evalutating machinability. Trans ASME 67:225–232
20. Komanduri R, Hou ZB (2001) A review of the experimental techniques for the measurement of heat and temperatures generated in some manufacturing processes and tribology. Tribol Int 34:653–682
21. Schmidt AO, Roubik JR (1949) Distribution of heat generated in drilling. Trans ASME 7:242–245
22. Shaw MC (1984) Metal cuting principles. Oxford Science Publications, Oxford
23. Kronenberg M (1966) Machining science and application. Theory and practice for operation and development of machining processes. Pergamon Press, London
24. Klocke F (2011) Manufacturing processes 1: cutting. Springer-Verlag, Berlin
25. Grzesik G (2008) Advanced machining processes of metallic materials: theory, modelling and applications. Elsevier, Oxford
26. Parashar BSN, Mittal RK (2006) Elements of manufacturing process. Prenstice, New Deli
27. Shaw M (2004) Metal Cutting Principles, 2nd edn. Oxford University Press, Oxford
28. Silin SS (1979) Similarity methods in metal cutting (in Russian). Mashinostroenie, Moscow
29. Smart EF, Trent M (1975) Temperature distribution in tools used for cutting iron, titanium and nickel. Int J Prod Res 13:265–290
30. Bejan A (1993) Heat transfer. Wiley, New York
31. Oxley PLB (1989) Mechanics of machining: an analytical approach to assessing machinability. Wiley, New York
32. Taylor FW (1907) On the art of cutting metals. Trans ASME 28:70–350
33. Astakhov VP, Davim PJ (2008) Tools (geometry and material) and tool wear. In: Davim PJ (ed) Machining: fundamentals and recent advances. Springer, London
34. Shvets SV (2010) New calculation of cutting characteristics. Russ Eng Res 30:478–483
35. Merchant ME (1945) Mechanics of the metal cutting process. I. Orthogonal cutting and a type 2 chip. J Appl Phys 16:267–275
36. Armarego EJ, Brown RH (1969) The machining of metals. Prentice-Hall, New Jersey

37. Finnie I, Shaw MC (1956) The friction process in metal cutting. Trans ASME 77:1649–1657
38. Usui E, Takeyma H (1960) A photoelastic analysis of machining stresses. ASME J Eng Ind 81:303–308
39. Zhang J-H (1991) Theory and technique of precision cutting. Pergamon Press, Oxford
40. Takeyama H, Usui H (1958) The effect of tool-chip contact area in metal cutting. ASME J Eng Ind 79:1089–1096
41. Chao BT, Trigger KJ (1959) Controlled contact cutting tools. Trans ASME 81:139–151
42. Usui E, Shaw MC (1962) Free machining steel—IV: tools with reduced contact length. Trans ASME B84:89–101
43. Hoshi K, Usui E (1962) Wear characteristics of carbide tools with artificially controlled tool-chip contact length. In: Proceedings of the third international MTDR conference. Pergamon Press, New York
44. Usui E, Kikuchi K, Hoshi K (1964) The theory of plasticity applied to machining with cut-away tools. Trans ASME B86:95–104
45. Jawahir IS, Van Luttervelt CA (1993) Recent developments in chip control research and applications. CIRP Ann 42:659–693
46. Luttervelt CA, Childs THC, Jawahir IS, Klocke F, Venuvinod PK (1998) Present situation and future trends in modelling of machining operations. Progress report of the CIRP working group 'modelling of machining operations'. CIRP Ann 74:587–626
47. Karpat Y, Ozel T (2008) Analytical and thermal modeling of high-speed machining with chamfered tools. J Manuf Sci Eng 130:1–15
48. Sadic MI, Lindstrom B (1992) The role of tool-chip contact length in metal cutting. J Mater Process Technol 37:613–627
49. Sadic MI, Lindstrom B (1995) The effect of restricted contact length on tool performance. J Mater Process Technol 48:275–282
50. Rodrigues AR, Coelho RT (2007) Influence of the tool edge geometry on specific cutting energy at high-speed cutting. J Braz Soc Mech Sci Eng XXIX:279–283
51. Astakhov VP (2004) The assessment of cutting tool wear. Int J Mach Tools Manuf 44:637–647
52. Stenphenson DA, Agapiou JS (2006) Metal cutting theory and practice, 2nd edn. CRC Press, Boca Raton
53. Byesrs JP (ed) (2006) Metalworking fluids, 2nd edn. CRC Press, Boca Raton
54. Silliman JD (ed) (1994) Cutting and grinding fluids: selection and application. SME, Dearborn
55. Jiang Z, Shang C (2010) Study on the mechanism of the cobalt leaching of cemented carbide in triethanolamine solution. Adv Mater Res 97–101:1203–1206
56. Xiaoming J, Xiuling Z, Suoxia H (2011) Study of composite inhibitor on the cobalt leaching of the cemented carbide tool. Adv Sci Lett 4:1256–1352
57. Bushman B, Gupta BK (1991) Handbook of tribology-materials, coatings, and surface treatments. McGraw-Hill, New York
58. Shaffer WR (1999) Cutting tool edge preparation. SME paper MR99-235
59. Byers JP (ed) (1994) Metalworking fluids. Marcel Dekker, New York
60. Kumabe J (1979) Vibration cutting. Jikkyo Publisher, Tokyo
61. Davis JR (ed) (1995) Tool materials (ASM specialty handbook). ASM, Metals Park
62. Isakov E (2000) Mechanical properties of work materials. Hanser Gardener Publications, Cincinnati
63. HSS Smart Guide, International High Speed Steel Research Forum. http://www.hssforum.com/. Accessed 15 Feb 2012
64. German RM, Smid I, Campbell LG, Keane J, Toth R (2005) Liquid phase sintering of tough coated hard particles. Int J Refract Metal Hard Mater 23:267–272
65. Astakhov VP, Joksch S (eds) (2012) Metal working fluids for cutting and grinding: fundamentals and recent advances. Woodhead, Cambridge
66. Pigott RJS, Colwell AT (1952) Hi-jet system for increasing tool life. SAE Q Trans 6:547–558

67. Wertheim R, Rotberg J, Ber A (1992) Influence of high-pressure flushing through the rake face of the cutting tool. CIRP Ann 41:101–106
68. Kovacevic R, Cherukuthota C, Mazurkewicz M (1995) High pressure watrejet cooling/lubrication to improve machining efficiency in milling. Int J Mach Tools Manuf 35:1458–1473
69. Lopez de Lacalle LN, Perez-Bilbatua J, Sanchez JA, Llorente JI, Gutierrez A, Alboniga J (2000) Using high pressure coolant in the drilling and turning of low machinability alloys. Int J Adv Manuf Technol 16:85–91
70. Kumar AS, Rahman M, Ng SL (2002) Effect of high-pressure coolant on machining performance. Int J Adv Manuf Technol 20:83–91
71. Ezugwu EO, Bonney J (2004) Effect of high-pressure coolant supply when machining nickel-base, Inconel 718, alloy with coated carbide tools. J Mater Process Technol 152–154: 1045–1050
72. Ezugwu EO, Da Silva RB, Bonney J, Machado AR (2205) Evaluation of the performance of CBN tools when turning Ti–6Al–4 V alloy with high pressure coolant supplies. Int J Mach Tools Manuf 45:1009–1014
73. Diniz AE, Micaroni R (2007) Influence of the direction and flow rate of the cutting fluid on tool life in turning process of AISI 1045 steel. Int J Mach Tools Manuf 47:247–254
74. Kamruzzaman M, Dhar NR (2009) Effect of high-pressure coolant on temperature, chip, force, tool wear, tool life and surface roughness in turning AISI 1060 steel. G.U. J Sci 22:359–370
75. Nandy AK, Gowrishankar MC, Paul S (2009) Some studies on high-pressure cooling in turning of Ti–6Al–4 V. Int J Mach Tools Manuf 49:182–198
76. Sharma VS, Dogra M, Suri NM (2009) Cooling techniques for improved productivity in turning. Int J Mach Tools Manuf 49:435–453
77. Kramar D, Krajnik P, Kopac J (2010) Capability of high pressure cooling in the turning of surface hardened piston rods. J Mater Process Technol 210:212–218
78. Mazurkiewicz M, Kubala Z, Chow J (1989) Metal machining with high-pressure water-jet cooling assistance—a new possibility. ASME J Eng Ind 111:7–12
79. Courbon C, Kramar D, Krajnik P, Pusavec F, Rech J, Kopa J (2009) Investigation of machining performance in high-pressure jet assisted turning of Inconel 718: an experimental study. Int J Mach Tools Manuf 49:1114–1125
80. Nakayama K, Arai M (1992) Comprehensive chip form classification based on the cutting mechanism. CIRP Ann 71:71–74
81. Nakayama K (1984) Chip control in metal cutting. Bull Jpn Soc Precis Eng 18:97–103
82. Astakhov VP (2010) Surface integrity definitions and importance in functional performance. In: Davim JP (ed) Surface integrity in machining. Springer-Verlag, London
83. Ezugwu EO (2005) High speed machining of aero-engine alloys. J Braz Soc Mech Sci Eng 26:1–11

Chapter 2
Tribology of Machining

M. J. Jackson, M. Whitfield, G. M. Robinson, J. Morrell and J. P. Davim

Abstract The chapter focuses on the tribological interactions in the area of machining and grinding and details the frictional interactions at the chip-tool interface. The chapter then discusses cutting, ploughing and sliding interaction models that are generally applied to the actions of cutting tools and provides solutions to reducing these interactions using solid and liquid lubricants. The authors intend to provide an understanding to readers of the chapter to design better processes for machining difficult-to-machine materials used in industry at large.

2.1 Frictional Interactions in Machining

Initial studies on chip-tool interactions during machining operations were carried out by Professor David Tabor and his team at the Cavendish Laboratory at the University of Cambridge in the United Kingdom during the late 1970 s. In their initial studies, Doyle et al. [17] constructed a transparent sapphire cutting tool bonded to a tool holder that transmitted the action of chip formation so that it could

M. J. Jackson (✉) · M. Whitfield · G. M. Robinson
Center for Advanced Manufacturing, Purdue University, West Lafayette, Indiana, USA
e-mail: bondedabrasive@gmail.com

J. Morrell
Y-12 National Security Complex, Oak Ridge, Tennessee, USA
e-mail: morrelljs@y12.doe.gov

J. P. Davim
Department of Mechanical Engineering, University of Aveiro,
Campus Santiago, 3810-193 Aveiro, Portugal
e-mail: pdavim@ua.pt

J. P. Davim (ed.), *Tribology in Manufacturing Technology*, Materials Forming, Machining and Tribology, DOI: 10.1007/978-3-642-31683-8_2,

be observed. The reflection of the freshly cut chip is transmitted through the tool by reflecting the image on to a projection face that is highly polished. Doyle et al. [17] reported that they used pure lead and pure tin in air to witness the mechanism of metal transfer to the cutting tool and defined the nature of contact in terms of contact zones as the chip moved across the surface of the tool at low cutting speeds. Initially, two zones were noted, one of sliding across the rake face of the tool (zone 1) and one consisting of the chip material sticking to the rake face (zone 2) ahead of zone 1. On further inspection of the images obtained using a cine camera, zone 1 comprised of two sub zones, namely: zone 1a (where the chip material slides at the edge of the cutting tool on its rake face) and zone 1b (where the chip material sticks to the cutting edge). Further studies by Horne et al. [27], further characterized the nature of contact between the cut chip and the surface of the tool. In an effort to understand the mechanics of chip formation and lubrication, a series of experiments were developed to understand how cooling lubricants provide a thin film between the chip and tool material. In their studies, various lubricants were used and dropped into the chip-tool zone in an effort to provide the means of separation between tool and chip. The lubricant was shown to enter the chip at the side of the material and is absorbed beneath the chip as the chip moves across the rake face. The formation of bubbles is also noticeable when machining aluminium with CCl_4, and may significantly contribute to the change in the mechanism of material transfer from chip to tool or the mechanics of machining. The work currently performed at Purdue University is based on work that was previously conducted at the Cavendish Laboratory and focuses on quantifying the interactions between chip and tool. This work investigates the applicability of applying orthogonal and oblique cutting theories. The initial work was completed by Madhavan [44]. Again, the use of transparent sapphire tools was employed. The cutting tool has a highly polished surface and was used with a research apparatus that is similar in action to the configuration of a metal planer. A Newport slide, model number PM500-4, was used for the linear slide with which to move the workpiece toward the cutting tool. The slide is controlled by a microprocessor. The velocity of the workpiece ranged from 25 mm/s to 150 mm/s. The use of an angle plate was necessary to attach the workpiece to the slide. Another slide was used to adjust the depth of cut of the cutting tool. The entire apparatus was mounted to a vibration isolation table. Soda-lime glass tool was used for comparison with the sapphire tool in order to allow for a comparison of frictional constants and sliding mechanisms between the two substrates. These were again highly polished using small cuboids of soda-lime glass for the rough polish. The final polish involved the use of 1 μm diameter cerium oxide particles. The experiments were imaged using an Olympus model OM-4T microscope. The magnification ranged from 50X to 200X. Force measurements were made with a Kistler piezoelectric transducer. The transducer interestingly enough was configured between the back of the cutting tool and the actual tool holder. The signal produced by the transducer was recorded using an oscilloscope. This was capable of recording the cutting force and the thrust force during machining.

These experiments were conducted for both smooth, polished tools and roughened tools. This was performed to determine if the tool would replicate the rake face of a conventional tool. The tools reflect the image of the rake face to the side of the tool for easy observation as described by Doyle et al. [17]. The initial experiments involved the cutting of wax to develop a basic idea of what may occur. An observation of the process showed that microcracks were formed during the planing of wax. This in turn created a crack along the shear plane. This observation would begin the basis of machining other plastically deformable materials in order to observe the mechanism of machining.

The use of pure lead was considered essential for the experiments concerned with understanding rake face interactions. This was performed using the dry cutting and lubricated conditions. The lubricant was composed of a mixture of two parts of oil and one part of ink. From these experiments, it was determined that three distinct zones form. The three zones are: Ia, Ib, and II. In zone Ia, the chip interacts with the tool that is known as intimate sliding contact. In zones Ib and II, the chip experiences sticking and metal transfer to the rake face of the tool. When the lubricant is used, metal transfer does not occur in zones Ib and II. It was observed by Doyle et alia [17], that zone II does not form when the experiment is conducted in a vacuum. The idea that there are three distinct zones is puzzling if two of the zones experience the same sliding action. Perhaps two zones with one zone composed of two sub zones would have provided a better explanation of the situation. The magnitude of shear strain is briefly mentioned. The average shear strain for lead is 40–100. The shear strain for aluminium and copper is approximately 20. There were no calculations included for these values, nor a discussion about the effect of strain causing chip curl. Further experiments were conducted in the later 1990s by Ackroyd [1], who used a new piece of apparatus for characterizing chip-tool interactions. This work further investigated the frictional interactions initially noted by Madhavan [44]. The cutting tool used was again a highly polished sapphire tool bit.

The experimental workpieces used included pure lead as well as pure tin. Both of these materials were obtained from Goodfellow in England. The percentage purity of both materials was 99.95 %. The dimensions of the workpiece were 50 mm square × 2 mm thick. Ackroyd [1] also investigated the machining of brass and pure aluminium. Sapphire tools were compared to aluminium and high-speed steel tools. With the use of these tools, a comparative analysis was undertaken to determine if the frictional effects were similar for all tool materials. The sapphire tool was once again polished with cerium oxide and then cleaned with acetone. The rake angles of the tools were 10° and −5°. The experimental apparatus uses a linear drive made by Anorad (model number LW5-750). This unit is capable of a maximum velocity of 2 m/s. The workpiece is similarly mounted to this unit to achieve the maximum workpiece velocity. The vertical stage used was the Newport 433 model. This stage is equipped with the Newport DMH-1, which is a digital micrometer. This allows for measurement of a 1 μm resolution over a 15 mm range. The Newport stage is attached to a Kistler 9254 dynamometer. This is attached to cast iron v-blocks that have been mounted to a worktable. The

Kistler unit is used in conjunction with a dual mode amplifier (Kistler 5010B). The optics used for observation can magnify up to 200X. The CCD camera used is a Sony DXC-930. The camera was used in conjunction with s-VHS video. The model used was a Panasonic AG-1970 with a capability of 30 frames per second. The digital high-speed camera used was a Kodak Motion Corder Analyzer Sr-Ultra. It is difficult to determine more than two zones resulting from the machining experiment. For the initial experiments, the following are the parameters were used: The depth of cut was 200 μm; width of cut was 2 mm; length of cut was 50 mm, and the cutting speed ranged from 0.5 mm/sec to 500 mm/sec. It was discovered, and was confirmed by Robinson [56], that the cutting edge plays a significant role in the machining experiments. As the nose radius becomes larger, the depth of cut determines whether the cutting tool cuts, or shears the workpiece material. A large cutting nose radius can create a negative rake angle on the cutting tool if the depth of cut is smaller than the nose radius.

From this work, it can be seen that there are two zones. The first zone involves the chip sliding with no deposit of metal. The second zone involves the chip sticking with some material deposited onto the tool. It was observed that a blunt tool has a higher normal force than a sharp tool. The frictional force versus normal force ratio decreases with the increase in velocity. It was also noted that there were no material deposits to the rake face at higher velocities. The discussion and conclusions of this work show that continual frictional forces increase. This is not a claim that can be substantiated. If frictional force continually increases with friction, it would reach a catastrophic point where the tool would fail. The conclusion was made using a single machining force measurement to examine both the sticking and sliding regions in question. Both zones cannot be observed simultaneously, therefore a direct comparison cannot be made.

Subsequent work completed by Hwang [29] followed closely the work of Ackroyd [1]. The apparatus that was previously used was upgraded with new components. The use of sapphire and high-speed tools for machining was investigated further by Hwang [29]. The experiments were designed to investigate the sliding and sticking zones. The apparatus was upgraded with a new linear slide. The linear slide was ball screw driven and was supplied by Parker as model number ERB80-B02LAJX-GXS677-A96. This allows for a larger power motor to drive the linear slide. The increased power allowed a larger variety of materials to be investigated. The slide has a maximum speed of 750 mm/sec. Again, the vertical stage used was a Newport model number 433. This is actuated using the Newport digital micrometer model number DMH-1. The imaging system remained the same as that used by Ackroyd [1].

A wider variety of plastically deformed materials were used in the experiments. The list includes oxygen free, high conductivity copper (OFHC), cartridge brass, pure lead, Al 1100 aluminium, and Al 6061-T6 aluminium. Hardness values were taken for all materials except for pure lead. The following are the Vickers' hardness values: OFHC = 89.7 kg mm^{-2}, cartridge brass = 153 kg mm^{-2}, Al 1100 = 49.5 kg mm^{-2}, and Al 6061-T6 = 116 kg mm^{-2}. The workpiece specimens investigated closely were pure lead and Al 6061-T6. The depth of cut for the lead was 200 μm, while the depth of cut for the 6061 was 100 μm. The cutting velocity spanned from 0.5 mm/s to 500 mm/s. The exact speeds used were 0.5 mm/s, 5 mm/s, 50 mm/s, and 500 mm/s, respectively. These experiments looked at the effects of lubrication on the chip formation process. The use of lubricant was observed to reduce the length of chip-tool contact length. The investigation focused on the rake face's sliding and sticking regions. This is a continuation of the work performed by Ackroyd [1] and shows the variation in machining different materials. Another aspect of the chip formation process was closely examined. The secondary deformation in a chip was explored with the use of a quick-stop experiment. This is an experiment that stops the cutting tool about three-quarters of the way across the workpiece. This will allow for the examination of the primary shear zone and grain orientation in the chip compared to the substrate. Another segment of the research investigated the use of modulation during machining pure metals with lubricant. The experiment was conducted with a vertical slide as opposed to a horizontal slide. It was not directly observed that the cutting fluid penetrated into the intermittent contact zone during these experiments. This assumption is made due to chip debris remaining static and the elimination of the metal deposit zone. During modulation, frictional forces are much smaller. It was noted that cutting remains under high pressure in the gap ahead of the cutting tool and that the cutting fluid reduced the region of the chip-tool contact. In this region, the contact length reduction promotes a reduction in frictional force. The application of cutting fluid will cause the zone of metal deposits to move further away from the tool edge. This will prevent the expansion of the stagnant metal zone. This is beneficial to eliminate a partial frictional constant. However, it is questionable as to whether modulation is solely responsible for this action.

The measurement of the rake face temperature was performed using an infrared imaging system. The signal is reflected at the back of the tool in order to investigate the temperature of the rake face of the tool. The methods of calculating temperature developed by Rapier, Boothroyd, and Loewen and Shaw were used to calculate the rake face temperature. These calculations were then compared to the measured data. From these results, it can be reasoned that modulation does not solve the frictional contact problem. However, modulation does create a constant cutting length and constant values of friction between chip and tool. However, this action alone is dependent upon the stroke length chosen, and will be different for each material machined and the corresponding depth of cut.

The research conducted by Lee [40] follows closely the work of both Ackroyd [1] and Hwang [29]. Lee's study included using particle image velocimetry to measure chip velocity. This utilized a linear slide with a ball screw. This allows for an adjustment of the depth of cut as little as 1 μm. A charge-coupled device (CCD) was used for high-speed imaging. Similar to the Hwang's work, a Kodak Motion Coder Analyzer Sr-Ultra is used. An optical microscope records the chip formation process, and employs a Nikon Optiphot that can examine the specimen up to 200X magnification. The CCD can capture images of up to 10,000 frames per second using a black and white format. The spatial resolution is 3.3 μm pixel size. The experiments were conducted dry. The workpiece specimens used were commercially pure lead and copper that is 99.95 % pure. These were both obtained from Goodfellow, UK. In the initial investigation using 6061-T6 aluminium, a built-up edge (BUE) was observed on the rake face. The BUE changes the rake angle thus increasing the chip velocity more than in the machining of other materials.

The velocity of the chip was inspected using constrained workpiece specimens. These areas show the rake face and the side view of the cut chip. This allows for the investigation of the metal deposit on the rake face and the effects of deformation in the secondary zone. A vertical stage moves the tool into contact with the workpiece. This is made possible through the use of a micrometer. At the foundation is the ball screw drive that brings the workpiece into contact with the cutting tool. This is a satisfactory experimental apparatus. However, there remains much to be desired with this particular apparatus in terms of rigidity and range of workpiece cutting speeds.

2.2 Cutting, Ploughing and Sliding Interactions (after [31])

The nature of the contact between surfaces is an important aspect of understanding the function of tribology in machining [31] and hereafter). The properties of the materials in contact are homogeneous and isotropic. Macrocontact conditions are most useful in models for friction when there is lubrication and the effects of surface heterogeneities are of little importance. Hertz's equations allow engineers to calculate the maximum compressive contact stresses and contact dimensions for non-conforming bodies in elastic contact. The parameters required to calculate the quantities and the algebraic equations used for simple geometries are given in Table 2.1 [65]. It should be noted that Hertz's equations apply to static, or quasi-static, elastic cases. In the case of sliding, plastic deformation, contact of very rough surfaces, or significant fracture, both the distribution of stresses and the contact geometry will be altered. Hertz's contact equations have been used in a range of component design applications, and in friction and wear models in which the individual asperities are modeled as simple geometric contacts.

Greenwood and Williamson [20] developed a surface geometry model that modeled contacts as being composed of a distribution of asperities. From that assumption, contact between such a surface and a smooth, rigid plane could be determined by three parameters: the asperity radius (R), the standard deviation of asperity heights (σ^*), and the number of asperities per unit area. To predict the extent of the plastic deformation of asperities, the plasticity index (ψ), also a function of the hardness (H), elastic modulus (E), and Poisson's ratio (ν), was introduced.

$$\psi = \left(\frac{E'}{H}\right)\left(\frac{\sigma *}{R}\right)^{1/2} \tag{2.1}$$

Table 2.1 Equations for calculating elastic (Hertz) contact stress [31] (Reproduced with permission)

Geometry	Contact dimension	Contact stress
Sphere-on-flat	$a = 0.721\sqrt[3]{PDE^*}$	$S_c = 0.918\sqrt[3]{P/(D^2E^*)^2}$
Cylinder-on-flat	$b = 1.6\sqrt{pDE^*}$	$S_c = 0.798\sqrt{p/DE^*}$
Cylinder-on-cylinder (axes parallel)	$b = 1.6\sqrt{pE^*/A}$	$S_c = 0.798\sqrt{pA/E^*}$
Sphere in a spherical socket	$a = 0.721\sqrt[3]{PE^*/B}$	$S_c = 0.918\sqrt{[3]P(B/E^*)^2}$
Cylinder in a circular groove	$b = 1.6\sqrt{pE^*/B}$	$S_c = 0.798\sqrt{pB/E^*}$

Symbol	Definition
P	Normal force
p	Normal force per unit contact length
$E_{1,2}$	Modulus of elasticity for bodies 1 and 2, respectively
$\nu_{1,2}$	Poisson's ratios for bodies 1 and 2, respectively
D	Diameter of the curved body, if only one if curved
$D_{1,2}$	Diameters of bodies 1 and 2, where $D_1 > D_2$ by convention
S_c	Maximum compressive stress
a	Radius of the elastic contact
b	Width of a contact (for cylinders)
E*	Composite modulus of bodies 1 and 2
A, B	Functions of the diameters of bodies 1 and 2

Where, $E' = E/(1 - \nu^2)$. This basic formulation was refined by various investigators such as Whitehouse and Archard [64] to incorporate other forms of height distributions, and the incorporation of a distribution of asperity radii, represented by the correlation distance β^*, which produced higher contact pressures and increased plastic flow. Therefore,

$$\psi = \left(\frac{E'}{H}\right)\left(\frac{\sigma^*}{\beta}\right)^{1/2} \tag{2.2}$$

Hirst and Hollander [25] used the plasticity index to develop diagrams to predict the start of scuffing wear. Other parameters, such as the average or root mean square slope of asperities, have been incorporated into wear models to account for such peculiarities [43]. Worn surfaces are observed to be much more complex than simple arrangements of spheres, or spheres resting on flat planes, and Greenwood readily acknowledged some of the problems associated with simplifying assumptions about surface roughness [21]. A comprehensive review of surface texture measurement methods have been given by Song and Vorburger [59]. The most commonly used roughness parameters are listed in Table 2.2. Parameters such as skewness are useful for determining lubricant retention qualities of surfaces, since they reflect the presence of cavities. However, one parameter alone cannot precisely model the geometry of surfaces. It is possible to have the same average roughness (or RMS roughness) for two different surfaces.

Table 2.2 Definitions of surface roughness parameters [31] (Reproduced with permission)

Let y_i = vertical distance from the ith point on the surface profile to the mean line
N = number of points measured along the surface profile
Thus, the following are defined:

Arithmetic average roughness	$R_a = \frac{1}{N}\sum_{i=1}^{N} \lvert y_i \rvert$
Root-mean-square roughness	$R_q = \left[\frac{1}{N}\sum_{i=1}^{N} y_i^2\right]^{1/2}$
Skewness	$R_{sk} = \frac{1}{NR_q^3}\sum_{i=1}^{N} y_i^3$ A measure of the symmetry of the profile $R_{sk} = 0$ for a Gaussian height distribution
Kurtosis	$R_{kurtosis} = \frac{1}{NR_q^4}\sum_{i=1}^{N} y_i^4$ A measure of the sharpness of the profile $R_{kurtosis} = 3.0$ for a Gaussian height distribution $R_{kurtosis} < 3.0$ for a broad distribution of heights $R_{kurtosis} > 3.0$ for a sharply-peaked distribution

Small amounts of wear can change the roughness of surfaces on the microscale and disrupt the nanoscale structure as well. Some of the following quantities have been used in models for friction:

1. The true area of contact;
2. The number of instantaneous contacts comprising the true area of contact;
3. The typical shapes of contacts (under load);
4. The arrangement of contacts within the nominal area of contact; and
5. The time needed to create new points of contact.

Finally, contact geometry-based models for friction generally assume that the normal load is constant. This assumption may be unjustified, especially when sliding speeds are relatively high, or when there are significant friction and vibration interactions in the tribosystem. As the sliding speed increases, frictional heating increases and surface thermal expansion can cause intermittent contact. The growth and excessive wear of intermittent contact points is termed thermo-elastic instability (TEI) [12]. TEI is only one potential source of the interfacial dynamics responsible for stimulating vibrations and normal force variations in sliding contacts. Another major cause is the eccentricity of rotating shafts, run-out, and the transmission of external vibrations. Static friction and stick–slip behavior are considered, and as with kinetic friction, the causes for such phenomena can be interpreted on several scales.

2.2.1 Static Friction and Stick–slip Phenomena

If all possible causes for friction are to be considered, it is reasonable to find out whether there are other means to cause bodies to stay together without the requirement for molecular bonding. Surfaces may adhere, but adherence is not identical to adhesion, because there is no requirement for molecular bonding. If a certain material is cast between two surfaces and, after penetrating and filling irregular voids in the two surfaces, solidifies to form a network of interlocking contacting points there may be strong mechanical joint produced, but no adhesion. Adhesion (i.e., electrostatically balanced attraction/chemical bonding) in friction theory meets the need for an explanation of how one body can transfer shear forces to another. Clearly, it is convenient to assume that molecular attraction is strong enough to allow the transfer of force between bodies, and in fact this assumption has led to many of the most widely used friction theories. From another perspective, is it not equally valid to consider that if one pushes two rough bodies together so that asperities penetrate, and then attempts to move those bodies tangentially, the atoms may approach each other closely enough to repel strongly, thus causing a backlash against the bulk materials and away from the interface. The repulsive force parallel to the sliding direction must be overcome to move the bodies tangentially, whether accommodation occurs by asperities climbing over one another, or by deforming one another. In the latter, it is repulsive forces and not adhesive bonding that produces sliding resistance. This section focuses on static friction and stick slip phenomena.

Ferrante et al. [19] have provided a comprehensive review of the subject. A discussion of adhesion and its relationship to friction has been conducted by Buckley [11]. Atomic probe microscopes permit investigators to study adhesion and lateral forces between surfaces on the atomic scale. The force required to shift the two bodies tangentially must overcome bonds holding the surfaces together. In the case of dissimilar metals with a strong bonding preference, the shear strength of the interfacial bonds can exceed the shear strength of the weaker of the two metals, and the static friction force (F_s) will depend on the shear strength of the weaker material (τ_m) and the area of contact (A). In terms of the static friction coefficient μ_s,

$$\mathbf{F}_s = \mu_s \mathbf{P}^* = \tau_m A \tag{2.3}$$

Or,

$$\mu_s = (\tau_m / \mathbf{P}^*) A \tag{2.4}$$

where $\mathbf{P}^*$, the normal force is comprised of the applied load and the adhesive contribution normal to the interface. Under specially controlled conditions, such as friction experiments with clean surfaces in vacuum, the static friction coefficients can be greater than 1.0, and the experiment becomes a test of the shear strength of the solid materials than of interfacial friction. Scientific understanding and approaches to modeling friction has been strongly influenced by concepts of solid surfaces and by the instruments available to study them. Atomic-force microscopes and scanning tunneling microscopes permit views of surface atoms with high resolution and detail. Among the first to study nanocontact frictional phenomena were McClellan et al. [41, 42]. A tungsten wire with a very fine tip is brought down to the surface of a highly oriented, cleaved basal plane of pyrolytic graphite as the specimen is oscillated at 10 Hz using a piezoelectric driver system. The cantilevered wire is calibrated so that its spring constant is known (2500 N/m) and the normal force could be determined by measuring the deflection of the tip using a reflected laser beam. As the normal force is decreased, the contributions of individual atoms to the tangential force became apparent. At the same time, it appeared that the motion of the tip became less uniform, exhibiting atomic-scale stick–slip.

Thompson and Robbins [61] discussed the origins of nanocontact stick–slip when analyzing the behavior of molecularly thin fluid films trapped between flat surfaces of face-centered cubic solids. At that scale, stick–slip was believed to arise from the periodic phase transitions between ordered static and disordered kinetic states. Immediately adjacent to the surface of the solid, the fluid assumed a regular, crystalline structure, but this was disrupted during each slip event. The experimental data points of friction force per unit area versus time exhibited extremely uniform classical stick–slip appearance. Once slip occurred, all the kinetic energy must be converted into potential energy in the film. In subsequent papers (Robbins et al. [54, 55] this group of authors used this argument to calculate the critical velocity, v_c, below which the stick–slip occurs is:

$$v_c = c(\sigma \mathbf{F}_s/M)^{1/2} \tag{2.5}$$

where σ is the lattice constant of the wall, $\mathbf{F}_s$ is the static friction force, M is the mass of the moving wall, and c is a constant.

Friction is defined as the resistance to relative motion between two contacting bodies parallel to a surface that separates them. Motion at the atomic scale is unsteady. In nanocontact, accounting for the tangential components of thermal vibrations of the atoms thus affects our ability to clearly define relative motion between surfaces. Under some conditions it may be possible to translate the surface laterally while the adhesive force between the probe tip and the opposite surface exceeds the externally applied tensile force. Landman et al. [39] reviewed progress in the field of molecular dynamics (MD). By conducting MD simulations of nickel rubbing a flat gold surface, Landman illustrated how the tip can attract atoms from the surface simply by close approach without actual indentation. A connective neck or bridge of surface atoms was observed to form as the indenter was withdrawn. The neck can exert a force to counteract the withdrawal force on the tip, and the MD simulations clearly model transfer of material between opposing asperities under pristine surface conditions. Landman has subsequently conducted numerous other MD simulations, including complete indentation and indentation in the presence of organic species between the indenter and substrate. Belak and Stowers [4], using a

Table 2.3 Static Friction Coefficients for clean metals in helium gas at two temperatures [31] (Reproduced with Permission)

Static friction coefficient		
Material combination	300 K	80 K
Fe (99.9 %) on Fe (99.99 %)	1.09	1.04
Al (99 %) on Al (99 %)	1.62	1.60
Cu (99.95 %) on Cu (99.95 %)	1.76	1.70
Ni (99.95 %) on Ni (99.95 %)	2.11	2.00
Au (99.98 %) on Au (99.98 %)	1.88	1.77
Ni (99.95 %) on Cu (99.95 %)	2.34	2.35
Cu (99.95 %) on Ni (99.95 %)	0.85	0.85
Au (99.98 %) on Al (99 %)	1.42	1.50
Fe (99.9 %) on Cu (99.95 %)	1.99	2.03

material volume containing 43,440 atoms in 160 layers, simulated many of the deformational features associated with metals, such as edge dislocations, plastic zones, and point defect generation. Calculated shear stresses for a triangular indenter passing along the surface exhibited erratic behavior, not unlike that observed during metallic sliding under clean conditions. Pollock and Singer [48] compiled a series of papers on atomic-scale approaches to friction.

While MD simulations and atomic-scale experiments continue to provide fascinating insights into frictional behavior, under idealized conditions, most engineering tribosystems are non-uniform. Not only are surfaces not atomically flat, but the materials are not homogeneous, and surface films and contaminant particles of many kinds, much larger than the atomic scale, may influence interfacial behavior. Static friction coefficients measured experimentally under ambient or contaminated conditions probably will not assume the values obtained in controlled environments. In a series of carefully conducted experiments on the role of adsorbed oxygen and chlorine on the shear strength of metallic junctions, Wheeler [63] showed how, μ_s, can be reduced in the presence of adsorbed gases. On the other hand, static friction coefficients for pure, well-cleaned metal surfaces in the presence of non-reactive gases like He can be relatively high. It is interesting to note that the friction of copper on nickel and the friction of nickel on copper are quite different. This is not an error, but rather a demonstration of the fact that reversing the materials of the sliding specimen and the counterface surface can affect the measured friction, confirming the assertion that friction is a property of the tribosystem and not of the materials in contact. A cryotribometer was used to obtain the data in Table 2.3. The length of time that two solids are in contact can also affect the relative role that adhesion plays in establishing the value of the static friction coefficient. Two distinct possibilities can occur: (a) if the contact becomes contaminated with a lower shear-strength species, the friction will decline; and (b) if the contact is clean and a more tenacious interfacial bond develops, the static friction will tend to increase. Akhmatov [2] demonstrated that by using cleaved rock salt that the formation of surface films over time lowers static friction. The opposite effect has been demonstrated for metals. A first approximation of rising static friction behavior is given by,

$$\mu_{S(t)} = \mu_{S(t=\infty)} - \left[\mu_{S(t=\infty)} - \mu_{S(t=0)}\right] e^{-ut} \tag{2.6}$$

where, $\mu_s(t)$, is the current value of the static friction coefficient at time t, $\mu_s(t = \infty)$ is the limiting value of the static friction coefficient at long times, $\mu_s(t = 0)$ is the initial static friction coefficient, and u is a rate constant. In contrast to exponential dependence on time, Buckley showed that by using data for tests of single-crystal Au touching Cu-5 % Al alloy that junction growth can cause the adhesive force to increase linearly with time.

When materials are placed in intimate contact, it is not unexpected that the atoms on their surfaces will begin to interact. The degree of this interaction will depend on the contact pressure, temperature, and the degree of chemical reactivity that the species have for each other, hence, static friction can change with the duration of contact. Despite the two opposite dependencies of static friction on time of contact, observations are consistent from a thermodynamic standpoint. Systems tend toward the lowest energetic state. In the case of interfaces, this state can be achieved either by forming bonds between the solids, or by forming bonds with other species (adsorbates and films) in the interface. The former process tends to strengthen the shear strength of the system, and the latter tends to weaken it. Sikorski [58] reported the results of experiments designed to compare friction coefficients of metals with their coefficients of adhesion (defined as the ratio of the force needed to break the bond between two specimens to the force which initially compressed them together). Rabinowicz [51] conducted a series of simple, tilting-plane tests with milligram- to kilogram-sized specimens of a variety of metals. Results demonstrated the static friction coefficient to increase as slider weight (normal force) decreased. For metal couples such as Au/Rh, Au/Au, Au/Pd, Ag/Ag, and Ag/Au, as the normal force increased over about six orders of magnitude (1 mg–1 kg), the static friction coefficients tended to decrease by nearly one order of magnitude.

Under low contact pressures, surface chemistry effects can play a relatively large role in governing static friction behavior. However, under more severe contact conditions, such as extreme pressures and high temperatures, other factors, more directly related to bulk properties of the solids, dominate static friction behavior. When very high pressures and temperatures are applied to solid contacts, diffusion bonds or solid-state welds can form between solids, and the term static friction ceases to be applicable. Table 2.4 lists a series of reported static friction coefficients. Note that in certain cases, the table references list quite different values for these coefficients. The temperature of sliding contact can affect the static friction coefficient. This behavior was demonstrated for single crystal ceramics by Miyoshi and Buckley [45], who conducted static friction tests of pure iron sliding on cleaned {0001} crystal surfaces of silicon carbide in a vacuum (10^{-8} Pa). For both $< 1010 >$ and $< 1120 >$ sliding directions, the static friction coefficients remained about level (0.4 and 0.5, respectively) from room temperature up to about 400 °C; then they each rose by about 50 % as the temperature rose to 800 °C. The authors attributed this effect to increased adhesion and plastic flow.

Table 2.4 Static friction coefficients for metals and non-metals (Dry or unlubricated conditions) [31] (Reproduced with permission)

Material combination			
Fixed specimen	Moving specimen	μ_s	References
Metals and Alloys on Various Materials			
Aluminium	Aluminium	1.05	[66]
	Steel, mild	0.61	[66]
	Titanium	0.54	[68]
Al, 6061-T6	Al, 6061-T6	0.42	[69]
	Copper	0.28	[69]
	Steel, 1032	0.35	[69]
	Ti-6Al-4 V	0.34	[69]
Copper	Cast iron	1.05	[66]
Steel	Cast iron	0.4	[67]
Steel, hardened	Steel, hardened	0.78	[66]
	Babbitt	0.42, 0.70	[66]
	Graphite	0.21	[66]
Steel, mild	Steel, mild	0.74	[66]
	Lead	0.95	[66]
Steel, 1032	Aluminium	0.47	[69]
	Copper	0.32	[69]
	Steel, 1032	0.31	[69]
	Ti-6Al-4 V	0.36	[69]
Steel, stainless 304	Copper	0.33	[69]
Tin	Iron	0.55	[68]
	Tin	0.74	[68]
Titanium	Aluminium	0.54	[68]
	Titanium	0.55	[68]

The role of adsorbed films on static friction suggests that one effective strategy for alleviating or reducing static friction is to introduce a lubricant or other surface treatment to impede the formation of adhesive bonds between mating surfaces. Contamination of surfaces from exposure to the ambient environment performs essentially the same function, but is usually less reproducible. Campbell [14] demonstrated how the treatment of metallic surfaces by oxidation can reduce the static friction coefficient. Oxide films were produced by heating metals in air. Sulfide films were produced by immersing the metals in sodium sulfide solution.

Except for the film on steel, film thicknesses were estimated to be 100–200 nm. Results from ten experiments, using a three ball-on-flat plate apparatus, were averaged to obtain static friction coefficients. In addition to producing oxides and sulfides, Campbell also tested oxide and sulfide films with Acto oil. The results of this investigation are shown in Table 2.5. For copper, the static friction coefficient ($\mu_s = 1.21$, with no film) decreased when the sulfide film thickness was increased from 0 to about 300 nm, after which the static friction coefficient remained about constant at 0.66.

Table 2.5 Reduction of static friction by surface films [31] (Reproduced with permission)

Material combination	μ_s, No film	μ_s, Oxide film	μ_s, Sulfide film
Copper-on-copper	1.21	0.76	0.66
Steel on steel	0.78	0.27	0.39
Steel on steel	078	0.19[a]	0.16[a]

[a] film and oil

The extent to which the solid lubricant can reduce static friction may be dependent on temperature, as confirmed by Hardy's earlier studies on the static friction of palmitic acid films on quartz. Between 20 and 50 °C the static friction coefficient decreases until melting occurs, at which time the lubricant loses its effectiveness.

Stick-slip is often referred to as a relaxation–oscillation phenomenon, and consequently, some degree of elasticity is needed in the sliding contact in order for stick–slip to occur. Israelachvili [30] considered stick–slip on a molecular level, as measured with surface forces apparatus. He considers the order–disorder transformations described by Thompson and Robbins [54, 61] in terms of simulations. Most classical treatments of stick–slip take a mechanics approach, considering that the behavior in unlubricated solid sliding is caused by forming and breaking adhesive bonds.

Stick-slip behavior can be modeled in several ways. Generally, the system is represented schematically as a spring-loaded contact, sometimes including a dashpot element to account for viscoelastic response [46]. The effects of time-dependent material properties on stick–slip behavior of metals is provided by Kosterin and Kragelski [34] and Kragelski [35]. Bowden and Tabor's analysis [8] considers a free surface of inertial mass m being driven with a uniform speed v in the positive x direction against an elastic constant k. Then the instantaneous resisting force F over distance x equals—kx. With no damping of the resultant oscillation,

$$ma = -kx \tag{2.7}$$

where acceleration $a = (d^2x/dt^2)$. The frequency n of simple harmonic motion is given by

$$n = (1/2\pi)(k/m)^{1/2} \tag{2.8}$$

Under the influence of a load **P** (mass **W** acting downward with the help of gravity g), the static friction force $\mathbf{F}_s$ can be represented as

$$\mathbf{F}_s = \mu_s \mathbf{P} \tag{2.9}$$

In terms of the deflection at the point of slip (x),

$$x = \mathbf{F}_s/k \tag{2.10}$$

If the kinetic friction coefficient μ is assumed to be constant during slip, then

$$ma - \mu\mathbf{P} = -kx \tag{2.11}$$

Letting time = 0 at the point of slip (where $x = F_s/k$), and the forward velocity $v \ll$ the velocity of slip, then,

$$x = (\mathbf{P}/k)[(\mu_s - \mu)\cos\omega t + \mu] \tag{2.12}$$

where, $\omega = (k/m)0.^{1/2}$ In this case, the magnitude of slip, δ, is

$$\delta = [\mathbf{P}(2\mu_s - 2\mu)/k] \tag{2.13}$$

From this equation, the larger the μ relative to μ_s, the less the effects of stick–slip, and when they are equal, the sliding becomes completely steady. Kudinov and Tolstoy [37] derived a critical velocity above which stick–slip could be suppressed. This critical velocity v_c was directly proportional to the difference in the static and kinetic friction coefficients $\Delta\mu$ and inversely proportional to the square root of the product of the relative dissipation of energy during oscillation ($\psi = 4\pi\tau$), the stiffness of the system k, and the slider mass m. Thus,

$$v_c = \Delta\mu N/\sqrt{\psi km} \tag{2.14}$$

where N is the factor of safety. The authors report several characteristic values of $\Delta\mu$ for slideways on machine tools: cast iron on cast iron = 0.08, steel on cast iron = 0.05, bronze on cast iron = 0.02, and PTFE on cast iron = 0.04.

System resonance within limited stick–slip oscillation ranges was discussed by Bartenev and Lavrentev [3], who cited experiments in which an oscillating normal load was applied to a system in which stick–slip was occurring. The minimum in stick–slip amplitude and friction force occurred over a range of about 1.5–2.5 kHz, the approximate value predicted by $(1/2\pi)(k/m)^{-1/2}$. Rabinowicz [50] suggested two possible solutions:

1. Decrease the slip amplitude or slip velocity by increasing contact stiffness, increasing system damping, or increasing inertia; and
2. Lubricate or otherwise form a surface film to ensure a positive μ versus velocity relationship.

The latter solution requires that effective lubrication be maintained, and stick–slip can return if the lubricant becomes depleted. The fact that stick–slip is associated with a significant difference between static and kinetic friction coefficients suggests that strategies that lower the former or raise the latter can be equally effective.

2.2.2 *Sliding Friction*

Sliding friction plays a very important role in many manufacturing processes. Sliding friction models, other than empirical models, can generally be grouped into five categories:

1. Plowing and cutting-based models;
2. Adhesion, junction-growth, and shear models;
3. Single- and multiple-layer shear models;
4. Debris layer and transfer layer models; and
5. Molecular dynamics' models.

Each type of model was developed to explain frictional phenomena. Some of the models are based on observations that contact surfaces contain grooves that are suggestive of a dominant contribution from plowing. Single-layer models rely on a view of the interface showing flat surfaces separated by a layer whose shear strength controls friction. Some models involve combinations, such as adhesion plus plowing. Recent friction models contain molecular-level phenomena. Lubrication-oriented models and the debris-based models describe phenomena that take place in zone I, whereas most of the classical models for solid friction concern zone II phenomena. There are few models that take into account the effects of both the interfacial properties and the surrounding mechanical systems such as zone III models.

2.2.2.1 Models for Sliding Friction

Sliding friction models are summarized in this section of the chapter and fall into one, or more, of the five categories explained in the previous section.

(a) Plowing Model. Plowing models assume that the dominant contribution to friction is the energy required to displace material ahead of a rigid protuberance or protuberances moving along a surface. One of the simplest models for plowing is that of a rigid cone of slant angle θ plowing through a surface under a normal load **P** [50]. If we assign a groove width w (i.e., twice the radius r of the circular section of the penetrating cone at surface level), the triangular projected area, A_p, swept out as the cone moves along is as follows:

$$A_p = \frac{1}{2}w(r\tan\theta) = \frac{1}{2}(2r)(r\tan\theta) = r^2\tan\theta \tag{2.15}$$

The friction force $\mathbf{F}_p$ for this plowing contribution to sliding is found by multiplying the swept-out area by the compressive strength p. Thus, $\mathbf{F}_p = (r^2 \tan \theta)\, p$, and the friction coefficient, if this were the only contribution, is $\mu_p = \mathbf{F}_p/\mathbf{P}$. From the definition of the compressive strength p as force per unit area, we can write:

$$p = P/\pi r^2 \tag{2.16}$$

And

$$\mu_p = \mathbf{F}_p/\mathbf{P} = \left(r^2\tan\theta\right)p/\pi r^2 p = (\tan\theta)/\pi \tag{2.17}$$

Table 2.6 Estimates of the maximum plowing contribution to friction. [31] (Reproduced with permission)

Metal	Critical rake angle[a] (degrees)	μ_p
Aluminium	−5	0.03
Nickel	−5	0.03
Lead	−35	0.22
α-Brass	−35	0.22
Copper	−45	0.32

[a] For a cone, the absolute value of the critical rake angle is 90 minus angle θ

This expression can also be written in terms of the apex angle of the cone α (= 90 − θ):

$$\mu_p = (2\cot\alpha)/\pi \tag{2.18}$$

Note that the friction coefficient calculated is for the plowing of a hard asperity and is not necessarily the same as the friction coefficient of the material sliding along the sides of the conical surface. Table 2.6 shows the maximum plowing contribution to friction for various metals.

(b) Adhesion, Junction Growth, and Shear (AJS) Models. The AJS interpretations of friction are based on a scenario in which two rough surfaces are brought close together, causing the highest peaks (asperities) to touch. As the normal force increases, the contact area increases and the peaks are flattened. Asperity junctions grow until they are able to support the applied load. Adhesive bonds form at the contact points. When a tangential force is applied, the bonds must be broken, and overcoming the shear strength of the bonds results in the friction force. Early calculations comparing bond strengths to friction forces obtained in experiments raised questions as to the general validity of such models. Observations of material transfer and similar phenomena suggested that the adhesive bonds might be stronger than the softer of the two bonded materials, and that the shear strength of the softer material, not the bond strength, should be used in friction models.

Traditional friction models, largely developed for metal-on-metal sliding, have added the force contribution due to the shear of junctions to the contribution from plowing, giving the extended expression:

$$\mu = (\tau A_r)/P + (\tan\theta)/\pi \tag{2.19}$$

where A_r is the real area of contact and τ is the shear strength of the material being plowed. This type of expression has met with relatively widespread acceptance in the academic community and is often used as the basis for other sliding friction models. But if the tip of the cone wears down, three contributions to the plowing process can be identified: the force needed to displace material from in front of the cone, the friction force along the leading face of the cone (i.e., the component in the macroscopic sliding direction), and the friction associated with shear of the interface along the worn frustum of the cone. From this analysis, it is clear that friction on two scales is involved: the macroscopic friction force for the entire system, and the friction forces associated with the flow of material along the face of the cone and across its frustum. That situation is somewhat analogous to the interpretation of orthogonal cutting of metals in which the friction force of the chip moving up along the rake face of the tool and friction along the wear land are not in general the same as the cutting force for the tool as a whole [5]. Considering the three contributions to the friction of a flat-tipped cone gives

$$\mu = (\tau/\mathbf{P})(\pi r^2) + \mu_i \cos^2\theta + (\tan\theta)/\pi \tag{2.20}$$

where r is defined as the radius of the top of the worn cone and μ_i is the friction coefficient of the cone against the material flowing across its face. Eq. 2.20 helps explain why the friction coefficients for ceramics and metals sliding on faceted diamond films are 10 or more times higher than the friction coefficients reported for smooth surfaces of the same materials sliding against smooth surfaces of diamond (i.e., $\mu \gg \mu_i$). When the rake angle θ is small, $\cos^2\theta$ is close to 1.0, and the second term is only slightly less than μ_i (0.02–0.12 typically). If one assumes that the friction coefficient for the material sliding across the frustum of the cone is the same as that for sliding along its face (μ_i), then Eq. 2.20 can be re-written:

$$\mu = 2\mu_i + (\tan\theta)/\pi \tag{2.21}$$

Thus, implying that the friction coefficient for a rigid sliding cone is more than twice that for sliding a flat surface of the same two materials. It is interesting to note that Eq. 2.21 does not account for the depth of penetration, a factor that seems critical for accounting for the energy required to plow through the surface (displace the volume of material ahead of the slider), and at $\theta = 90^\circ$, which implies infinitely deep penetration of the cone, it would be impossible to move the slider at all as μ tends toward infinity.

When one the complexities of surface finish it seems remarkable that Eqs. 2.20 and 2.21, which depend on a single quantity [(tan θ)/π], should be able to predict the friction coefficient with any degree of accuracy. The model is based on a single conical asperity cutting through a surface that makes no obvious accountability for multiple contacts and differences in contact angle. The model is also based on a surface's relatively ductile response to a perfectly rigid asperity and can neither account for fracture during wear nor account for the change in the groove geometry that one would expect for multiple passes over the same surface.

Mulhearn and Samuels [47] published a paper on the transition between abrasive asperities cutting through a surface and plowing through it. The results of their experiments suggested that there exists a critical rake angle for that type of transition. (Note: The rake angle is the angle between the normal to the surface and the leading face of the asperity, with negative values indicating a tilt toward the direction of travel.) If plowing can occur only up to the critical rake angle, then we may compute the maximum contribution to friction due to plowing from the data of Mulhearn and Samuels and Eq. 2.18 (Table 2.6). This approach suggests that the maximum contribution of plowing to the friction coefficient of aluminium or nickel is about 0.03 in contrast to copper, whose maximum plowing contribution is 0.32.

Since the sliding friction coefficient for aluminium can be quite high (over 1.0 in some cases), the implication is that factors other than plowing, such as the shearing of strongly adhering junctions, would be the major contributor. Examination of unlubricated sliding wear surfaces of both Al and Cu often reveals a host of ductile-appearing features not in any way resembling cones, and despite the similar appearances in the microscope of worn Cu and Al, one finds from the first and last rows in Table 2.6 that the contribution of plowing to friction should be different by a factor of 10. Again, the simple cone model appears to be too simple to account for the difference.

Hokkirigawa and Kato [26] carried the analysis of abrasive contributions to sliding friction even further using observations of single hemispherical sliding contacts (quenched steel, tip radius 26 or 62 μm) on brass, carbon steel, and stainless steel in a scanning electron microscope. They identified three modes: (a) plowing (b) wedge formation and (c) cutting (chip formation). The tendency of the slider to produce the various modes was related to the degree of penetration, D_p. Here, D_p. Here, $D_p = h/a$, where h is the groove depth and a is the radius of the sliding contact. The sliding friction coefficient was modeled in three ways depending upon the regime of sliding. Three parameters were introduced: $f = p/\tau \quad \theta = \sin^{-1}(a/R)$

and β, the angle of the stress discontinuity (shear zone) from Challen and Oxley's [14] analysis. Where p is the contact pressure, τ is the bulk shear stress of the flat specimen, and R is the slider tip radius. The friction coefficient was given as follows for each mode:

Cutting mode:

$$\mu = \tan\left[\theta - (\pi/4) + \frac{1}{2}\cos^{-1} f\right] \tag{2.22}$$

Wedge-forming mode:

$$\mu = \frac{\left\{1 - \sin 2\beta + (1 - f^2)^{1/2}\right\} \sin\theta + f\cos\theta}{\left\{1 - \sin 2\beta + (1 - f^2)^{1/2}\right\} \cos\theta + f\sin\theta} \tag{2.23}$$

Plowing mode:

$$\mu = \frac{A\sin\theta + \cos(\cos^{-1} f - \theta)}{A\sin\theta + \cos(\cos^{-1} f - \theta)} \tag{2.24}$$

Where

$$A = 1 + (\pi/2) + \cos^{-1} f - 2\theta - 2\sin^{-1}\frac{\sin\theta}{(1 - f)^{-1/2}} \tag{2.25}$$

For unlubricated conditions, the transitions between the various modes were experimentally determined by observation in the scanning electron microscope.

Table 2.7 Critical degree of penetration (D_p) for unlubricated friction mode transitions [31] (Reproduced with permission)

Material	Value of D_p for the transition	
	Plowing to wedge formation	Wedge formation to cutting
Brass	0.17 (tip radius 62 μm)	0.23 (tip radius 62, 27 μm)
Carbon steel	0.12 (tip radius 62 μm)	0.23 (tip radius 27 μm)
Stainless steel	0.13 (tip radius 62, 27 μm)	0.26 (tip radius 27 μm)

Table 2.7 summarizes those results. Results of the study illustrate the point that the analytical form of the frictional dependence on the shape of asperities cannot ignore the mode of surface deformation. In summary, the foregoing treatments of the plowing contribution to friction assumed that asperities could be modeled as regular geometric shapes. However, rarely do such shapes appear on actual sliding surfaces. The asperities present on most sliding surfaces are irregular in shape, as viewed with a scanning electron microscope.

(c) Plowing with Debris Generation. Even when the predominant contribution to friction is initially from cutting and plowing of hard asperities through the surface, the generation of wear debris that submerges the asperities can reduce the severity of plowing. Table 2.8 shows that starting with multiple hard asperities of the same geometric characteristics produced different initial and steady-state friction coefficients for the three slider materials. Wear debris accumulation in the contact region affected the frictional behavior. In the case of abrasive papers and grinding wheels, this is called loading. Loading is extremely important in grinding, and a great deal of effort has been focused on dressing grinding wheels to improve their material removal efficiency. One measure of the need for grinding wheel dressing is an increase in the tangential grinding force or an increase in the power drawn by the grinding spindle.

As wear progresses, the wear debris accumulates between the asperities and alters the effectiveness of the cutting and plowing action by covering the active points. If the cone model is to be useful at all for other than pristine surfaces, the effective value of θ must be given as a function of time or number of sliding passes. Not only is the wear rate affected, but the presence of debris affects the interfacial shear strength, as is explained later in this chapter in regard to third-body particle effects on friction. The observation that wear debris can accumulate and so affect friction has led investigators to try patterning surfaces to create pockets where debris can be collected [60]. The orientation and depths of the ridges and grooves in a surface affect the effectiveness of the debris-trapping mechanism.

(d) Plowing with Adhesion. Traditional models for sliding friction have historically been developed with metallic materials in mind. Classically, the friction force is said to be an additive contribution of adhesive (S) and plowing forces (F_{pl}) [8]:

$$\mathbf{F} = \mathbf{S} + \mathbf{F}_{pl} \tag{2.26}$$

Table 2.8 Effects of material type on friction during abrasive sliding[a] [31] (Reproduced with permission)

	24 μm Grit size		16 μm Grit size	
Slider material	Starting μ	Ending μ	Starting μ	Ending μ
AISI 52100 steel	0.47	0.35	0.45	0.29
2014-T4 aluminium	0.69	0.56	0.64	0.62
PMMA	0.73	0.64	0.72	0.60

[a] Normal force 2.49 N, sliding speed 5 mm/sec, multiple strokes 20 mm long

The adhesive force derives from the shear strength of adhesive metallic junctions that are created when surfaces touch one another under a normal force. Thus, by dividing by the normal force we find that $\mu = \mu_{\text{adhesion}} + \mu_{\text{plowing}}$. If the shear strength of the junction is τ and the contact area is A, then

$$\mathbf{S} = \tau A \tag{2.27}$$

The plowing force F_{pl} is given by

$$\mathbf{F}_{pl} = pA' \tag{2.28}$$

Where p is the mean pressure to displace the metal in the surface and A' is the cross section of the grooved wear track. While helpful in understanding the results of experiments in the sliding friction of metals, the approach involves several applicability-limiting assumptions, for example, that adhesion between the surfaces results in bonds that are continually forming and breaking, that the protuberances of the harder of the two contacting surfaces remain perfectly rigid as they plow through the softer counterface, and perhaps most limiting of all, that the friction coefficient for a tribosystem is determined only from the shear strength properties of materials.

(e) Single-Layer Shear (SLS) Models. SLS models for friction depict an interface as a layer whose shear strength determines the friction force, and hence, the friction coefficient. The layer can be a separate film, like a solid lubricant, or simply the near surface zone of the softer material that is shearing during friction. The friction force **F** is the product of the contact area A and the shear strength of the layer:

$$\mathbf{S} = \tau A \tag{2.29}$$

The concept that the friction force is linearly related to the shear strength of the interfacial material has a number of useful implications, especially as regards the role of thin lubricating layers, including oxides and tarnish films. It is known from the work of Bridgman [9] on the effects of pressure on mechanical properties that τ is affected by contact pressure, p:

$$\tau = \tau_o + \alpha p \tag{2.30}$$

Table 2.9 Measured values for the shear stress dependence on pressure [31](Reproduced with permission)

Material	τ_o (kgf/mm^2)	α
Aluminium	3.00	0.043
Beryllium	0.45	0.250
Chromium	5.00	0.240
Copper	1.00	0.110
Lead	0.90	0.014
Platinum	9.50	0.100
Silver	6.50	0.090
Tin	1.25	0.012
Vanadium	1.80	0.250
Zinc	8.00	0.020

Table 2.9 lists several values for the shear stress and the constant α (Kragelskii et alia [36].

(f) Multiple-Layer Shear (MLS) Models. SLS models presume that the sliding friction can be explained on the basis of the shear strength on a single layer interposed between solid surfaces. Evidence revealed by the examination of frictional surfaces suggests that shear can occur at various positions in the interface: for example, at the upper interface between the solid and the debris layer, within the entrapped debris or transfer layer itself, at the lower interface, or even below the original surfaces where extended delaminations may occur. Therefore, one may construct a picture of sliding friction that involves a series of shear layers (sliding resistances) in parallel. Certainly, one would expect the predominant frictional contribution to be the lowest shear strength in the shear layers. Yet the shear forces transmitted across the weakest interface may still be sufficient to permit some displacement to occur at one or more of the other layers above or below it, particularly if the difference in shear strengths between those layers is small.

The MLS models can be treated like electrical resistances in a series. The overall resistance of such a circuit is less than any of the individual resistances because multiple current paths exist. Consider, for example, the case where there are three possible operable shear planes stacked up parallel to the sliding direction in the interface. Then,

$$\frac{1}{\mathbf{F}}=\frac{1}{\mathbf{F}_1}+\frac{1}{\mathbf{F}_2}+\frac{1}{\mathbf{F}_3} \tag{2.31}$$

And, solving for the total friction force F, in terms of the friction forces acting on the three layers, is

$$\mathbf{F}=\frac{\mathbf{F}_1\mathbf{F}_2\mathbf{F}_3}{\mathbf{F}_1\mathbf{F}_2+\mathbf{F}_2\mathbf{F}_3+\mathbf{F}_1\mathbf{F}_3} \tag{2.32}$$

If the area of contact A is the same across each layer, then Eq. 2.32 can be written in terms of the friction coefficient of the interface, the shear stresses of each layer, and the normal load **P** as follows:

$$\mu = \left(\frac{\mathbf{A}}{\mathbf{P}}\right)\left[\frac{\tau_1 \tau_2 \tau_3}{\tau_1 \tau_2 + \tau_2 \tau_3 + \tau_1 \tau_3}\right] \tag{2.33}$$

If one of the shear planes suddenly became unable to deform (say, by work hardening or by clogging with a compressed clump of wear debris), the location of the governing plane of shear may shift quickly, causing the friction to fluctuate. Thus, by writing the shear stresses of each layer as functions of time, the MLS model has the advantage of being able to account for variations in friction force with time and may account for some of the features observed in microscopic examinations of wear tracks.

(g) Molecular Dynamics' Models. When coupled with information from nanoprobe instruments, such as the atomic force microscope, the scanning tunneling microscope, the surface-forces apparatus, and the lateral-force microscope, MD studies have made possible insights into the behavior of pristine surfaces on the atomic scale. Molecular dynamics models of friction for assemblages of even a few hundred atoms tend to require millions upon millions of individual, iterative computations to predict frictional interactions taking place over only a fraction of a second in real time, because they begin with very specific arrangements of atoms, usually in single crystal form with a specific sliding orientation, results are often periodic with sliding distance. Some of the calculation results are remarkably similar to certain types of behavior observed in real materials, simulating such phenomena as dislocations (localized slip on preferred planes) and the adhesive transfer of material to the opposing counterface. However, molecular dynamics models are not presently capable of handling such contact surface features as surface fatigue-induced delaminations, wear debris particles compacting and deforming in the interface, high-strain-rate phenomena, work hardening of near-surface layers, and effects of inclusions and other artifacts present in the microstructures of commercial engineering materials.

The models presented up to this point use either interfacial geometric parameters or materials properties (i.e., bonding energies, shear strengths, or other mechanical properties) to predict friction. Clearly, frictional heating and the chemical environment may affect some of the variables used in these models. For example, the shear strength of many metals decreases as the temperature increases and increases as the speed of deformation increases. Certainly, wear and its consequences (debris) will affect friction. Thus, any of the previously described models will probably require some sort of modification, depending on the actual conditions of sliding contact. In general, the following can be said about friction models:

1. No existing friction model explicitly accounts for all the possible factors that can affect friction;
2. Even very simple friction models may work to some degree under well-defined, limited ranges of conditions, but their applicability must be tested in specific cases;
3. Accurately predictive, comprehensive tribosystem-level models that account for interface geometry, materials properties, lubrication aspects, thermal, chemical, and external mechanical system response, all in a time-dependent context, do not exist;

4. Friction models should be selected and used based on an understanding of their limitations and on as complete as possible an understanding of the dominant influences in the tribosystem to which the models will be applied; and
5. Current quantitative models produce a single value for the friction force, or friction coefficient. Since the friction force in nearly all known tribosystems varies to some degree, any model that predicts a single value is questionable.

If no existing model is deemed appropriate, the investigator could either modify a current model to account for the additional variables, develop a new system-specific model, or revert to simulative testing and/or field experiments to obtain the approximate value. An alternative to modeling is to estimate frictional behavior using a graphical, or statistical approach.

2.2.3 Frictional Heating

Heat generation and rising surface temperatures are intuitively associated with friction. When a friction force **F** moves through a distance x, an amount of energy $\mathbf{F}x$ is produced. The laws of thermodynamics require that the energy so produced be dissipated to the surroundings. At equilibrium, the energy into a system $\mathbf{U}_{in}$ equals the sum of the energy output to the surroundings $\mathbf{U}_{out}$ (dissipated externally) and the energy accumulated $\mathbf{U}_{accumulated}$ (consumed or stored internally):

$$\mathbf{U}_{in} = \mathbf{U}_{out} + \mathbf{U}_{accumulated} \tag{2.34}$$

The rate of energy input in friction is the product of F and the sliding velocity v whose units work out to energy per unit time (e.g., Nm/sec). This energy input rate at the frictional interface is balanced almost completely by heat conduction away from the interface, either into the contacting solids or by radiation or convection to the surroundings. In general, only a small amount of frictional energy, perhaps only 5 %, is consumed or stored in the material as microstructural defects such as dislocations, the energy to produce phase transformations, surface energy of new wear particles and propagating subsurface cracks, etc. Most of the frictional energy is dissipated as heat. Under certain conditions, there is enough heat to melt the sliding interface. Energy that cannot readily be conducted away from the interface raises the temperature locally. Assuming that the proportionality of friction force F to normal force **P** (i.e., by definition, $\mathbf{F} = \mu\mathbf{P}$) holds over a range of normal forces, we would expect that the temperature rise in a constant-velocity sliding system should increase linearly with the normal force. Tribologists distinguish between two temperatures, the flash temperature and the mean surface temperature. The former is localized, the latter averaged out over the nominal contact zone. Since sliding surfaces touch at only a few locations at any instant, the energy is concentrated there and the heating is particularly intense—thus, the name flash temperature. The combined effect of many such flashes dissipating their energy in the interface under steady state is to heat a near-surface layer to an average

Table 2.10 Temperature rise during sliding. after Jaeger [32, 31] (Reproduced with permission)

Conditions	Temperature Rise[a] ($T = T_o$)
Circular junction of radius a	$= \frac{Q}{4a(k_1+k_2)}$
Square junction of side = 2 *l*, at low speed	$= \frac{Q}{4.24l(k_1+k_2)}$
Square junction of side = 2 *l*, at high speed wherein the slider is being cooled by the incoming surface of the flat disk specimen	$= \frac{Qx^{1/2}}{3.76l\left[k_1(l/v)^{1/2}+1.125x^{1/2}k_2\right]}$ Where x = $(k_1/\rho_1 c_1)$ for the disk material

T = steady-state junction temperature, T_o = initial temperature, $k_{1,2}$ = thermal conductivity of the slider and flat bodies, ρ = density, c = specific heat

temperature that is determined by the energy transport conditions embodied in Eq. 5.34 given earlier. Blok [6] discussed the concept and calculation of flash temperature in a review article. The early work of Blok [7] and Jaeger [32] is still cited as a basis for more recent work, and it has been reviewed in a simplified form by Bowden and Tabor [8]. Basically, the temperature rise in the interface is given as a function of the total heat developed, Q:

$$Q = \frac{\mu \mathbf{W} g v}{J} \tag{2.35}$$

where μ is the sliding friction coefficient, **W** is the load, g the acceleration due to gravity, v the sliding velocity, and J the mechanical equivalent of heat (4.186 J/cal). Expressions for various heat flow conditions are then developed based on Eq. 2.35. Some of these are given in Table 2.10.

As Table 2.10 shows, the expressions become more complicated when the cooling effects of the incoming, cooler surface are accounted for. Rabinowicz [50] published an expression for estimating the flash temperature rise in sliding:

$$\theta_m = \frac{v}{2}(\pm \text{ a factor of 2 to 3}) \tag{2.36}$$

Where, v is sliding velocity (ft/min) and θ_m is the estimated surface flash temperature (°F). A comparison of the results of using Eq. 2.36 with several other, more complicated models for frictional heating has provided similar results, but more rigorous treatments are sometimes required to account for the variables left out of this rule of thumb. In general, nearly all models for flash or mean temperature rise during sliding contain the friction force–velocity product. Sometimes, the friction force is written as the product of the normal force and friction coefficients.

A review of frictional heating calculations has been provided by Cowan and Winer [15], along with representative materials properties data to be used in those calculations. Their approach involves the use of two heat partition coefficients (γ_1 and γ_2) that describe the relative fractions of the total heat that go into each of the contacting bodies, such that $\gamma_1 + \gamma_2 = 1$. The time that a surface is exposed to frictional heating will obviously affect the amount of heat it receives. The Fourier modulus, F_o, a dimensionless parameter, is introduced to establish whether or not steady-state conditions have been reached at each surface. For a contact radius a, an exposure time t, and a thermal diffusivity for body i of D_i,

$$F_o = \frac{D_i t}{a^2} \tag{2.37}$$

The Fourier modulus is taken to be 100 for a surface at steady state conditions. Another useful parameter grouping is the Peclet number P_e, defined in terms of the density of the solid ρ, the specific heat c_p, the sliding velocity v, the thermal conductivity k, and the characteristic length L_c:

$$P_e = \frac{\rho c_p v L_c}{k} \tag{2.38}$$

The characteristic length is the contact width for a line contact or the contact radius for a circular contact. The Peclet number relates the thermal energy removed by the surrounding medium to that conducted away from the region in which frictional energy is being dissipated. As $D_i = (\rho c_p/k)$ yields the following,

$$P_e = \frac{v L_c}{D_i} \tag{2.39}$$

The Peclet number is sometimes used as a criterion for determining when to apply various forms of frictional heating models. Peclet number is used in understanding frictional heating problems associated with grinding and machining processes. It is important to compare the forms of models derived by different authors for calculating flash temperature rise. Four treatments for a pin moving along a stationary flat specimen are briefly compared: Rabinowicz's derivation based on surface energy considerations, a single case from Cowan and Winer's review, Kuhlmann-Wilsdorf's model, and the model provided by Ashby. Based on considerations of junctions of radius r and surface energy of the softer material Γ, Rabinowicz arrived at the following expression:

$$T_f = \frac{3000\ \pi \mu \Gamma v}{J(k_1 + k_2)} \tag{2.40}$$

where J is the mechanical equivalent of heat, v is sliding velocity, μ is the friction coefficient, and k_1 and k_2 represent the thermal conductivities of the two bodies. The constant 3000 obtained from the calculation of the effective contact radius r in terms of the surface energy of the circular junctions Γ and their hardness H (i.e., $r = 12{,}000\Gamma H$) and the load carried by each asperity ($P = \pi r^2 H$). Thus, the numerator is actually the equivalent of Fv expressed in terms of the surface energy model. The equation provided by Cowan and Winer, for the case of a circular contact with one body in motion is

$$T_f = \frac{\gamma_1 \mu P v}{\pi a k_1} \tag{2.41}$$

Table 2.11 Effects of deformation type and peclet number on flash temperature calculation for the circular contact case [31] (Reproduced with permission)

Type of deformation	Peclet number	Average flash temperature[a]
Plastic	$P_e < .02$	$T_f = \mu \mathbf{P}^{0.5} v \frac{\sqrt{\pi\rho}}{8k}$
Plastic	$P_e > 200$	$T_f = 0.31 \mu \mathbf{P}^{0.25} v^{0.5} \left[\frac{(\pi\rho)^{0.75}}{(k\rho C)^{0.5}}\right]$
Elastic	$P_e < .02$	$T_f = 0.13 \mu \mathbf{P}^{0.667} v \left(\frac{1}{k}\right)\left(\frac{E_v}{R}\right)^{0.333}$
Elastic	$P_e > 200$	$T_f = 0.36 \mu \mathbf{P}^{0.5} v^{0.5} \left(\frac{1}{\sqrt{k\rho c}}\right)\left(\frac{E_v}{R}\right)^{0.5}$

[a] Key: μ = friction coefficient, **P** = load, v = velocity, k = thermal conductivity, π = density, c = heat capacity, E_v = the reduced elastic modulus = $E/(1-v^2)$, v = Poisson's ratio, ρ = flow pressure of the softer material

where γ_1 is the heat partition coefficient, described earlier, P is the normal force, a is the radius of contact, and k_1 is as defined earlier. The value of γ_1 takes various forms depending on the specific case. The presence of elastic, or plastic contact, can also affect the form of the average flash temperature, as Table 2.11 demonstrates. Here, the exponents of normal force and velocity are not unity in all cases. Kuhlmann-Wilsdorf [38] considered an elliptical contact area as the planar moving heat source. The flash temperature is given in terms of the average temperature in the interface T_{ave}:

$$T_{ave} = \frac{\pi q r}{4k_1} \tag{2.42}$$

Where q is the rate of heat input per unit area (related to the product of friction force and velocity), r is the contact spot radius, and k_1 is the thermal conductivity, as given earlier. Then

$$T_f = \frac{T_{avc}}{(1/ZS) + (k_1/S_O)} \tag{2.43}$$

where Z is a velocity function and S and S_o are contact area shape functions (both = 1.0 for circular contact). At low speeds, where the relative velocity of the surfaces $v_r < 2 (v_r = v/P_e)$, Z can be approximated by $1/[1 + (v_r/3)]$. The differences between models for frictional heating arise from the following:

1. Assuming different shapes for the heat source on the surface;
2. Different ways to partition the flow (dissipation) of heat between sliding bodies;
3. Different ways to account for thermal properties of materials (e.g., using thermal diffusivity instead of thermal conductivity, etc.);
4. Different contact geometry (sphere-on-plane, flat-on-flat, cylinder-on-flat, etc.);
5. Assuming heat is produced from a layer (volume) instead of a planar area; and
6. Changes in the form of the expression as the sliding velocity increases.

Comparing the temperature rises predicted by different models for low sliding speeds produces accurate results, even with the uncertainties in the values of the material properties that go into the calculations. At higher speeds, the predictions become unreliable since materials properties change as a function of temperature and the likelihood of the interface reaching a steady state is much lower. Experimental studies have provided very useful information in validating the forms of frictional heating models. Experimental scientists have often used embedded thermocouples in one or both members of the sliding contact to measure surface temperatures, and others sometimes made thermocouples out of the contacts themselves. However, techniques using infrared sensors have been used as well. Dow and Stockwell [16] used infrared detectors with a thin, transparent sapphire blade sliding on a 15-cm-diameter ground cylindrical drum to study the movements and temperatures of hot spots. Griffioen et al. [22] and Quinn and Winer [49] used an infrared technique with a sphere-on-transparent sapphire disk geometry. A similar arrangement was also developed and used by Furey with copper, iron, and silver spheres sliding on sapphire, and Enthoven et al. [18]. used an infrared system with a ball-on-flat arrangement to study the relationship between scuffing and the critical temperature for its onset.

Frictional heating is important because it changes the shear strengths of the materials in the sliding contact, promotes reactions of the sliding surfaces with chemical species in the environment, enhances diffusion of species, and can result in the breakdown or failure of the lubricant to perform its functions. Under extreme conditions, such as plastic extrusion, frictional heating can result in molten layer formation that serves as a liquid lubricant.

2.3 Lubrication to Control Friction in Machining (after Jackson and Morrell [31])

The frictional characteristics of liquid and solid lubricants and their interaction with materials are reviewed. Comprehensive discussions of the mechanical and chemical engineering aspects of lubrication are available in the literature [62]. The following section was originally published by Jackson and Morrell [31] and appears hereafter.

2.3.1 Liquid Lubrication

The process of lubrication is one of supporting the contact pressure between opposing surfaces, helping to separate them, and at the same time reducing the sliding or rolling resistance in the interface. There are several ways to accomplish this. One way is to create in the gap between the bodies geometric conditions that produce a fluid pressure sufficient to prevent the opposing asperities from touching while still permitting shear to be fully accommodated within the fluid. That method relies on fluid mechanics and modifications of the lubricant chemistry to tailor the liquid's properties. Another way to create favorable lubrication conditions is to formulate the liquid lubricant in such a way that chemical species within it react with the surface of the bodies to form shearable solid films. Surface species need not react with the lubricant, but catalyze the reactions that produce these protective films.

Several attributes of liquids make them either suitable or unsuitable as lubricants. Klaus and Tewksbury [33] have discussed these characteristics in some detail. They include:

1. Density;
2. Bulk modulus;
3. Gas solubility;
4. Foaming and air entrainment tendencies;
5. Viscosity and its relationships to temperature and pressure;
6. Vapor pressure;
7. Thermal properties and stability; and
8. Oxidation stability.

The viscosity of fluids usually decreases with temperature and therefore can reduce the usefulness of a lubricant as temperature rises. The term viscosity index, abbreviated VI, is a means to express this variation. The higher the VI, the less the change in viscosity with temperature. One of the types of additives used to reduce the sensitivity of lubricant viscosity to temperature changes is called a VI improver. ASTM test method D 2270 is one procedure used to calculate the VI. The process is described step-by-step in the article by Klaus and Tewksbury [33]. The method involves references to two test oils, the use of two different methods of calculation (depending on the magnitude of VI), and relies on charts and tables.

ASTM Standard D341 recommends using the Walther equation to represent the dependence of lubricant viscosity on temperature. Defining Z as the viscosity in cSt plus a constant (typically ranging from 0.6 to 0.8 with ASTM specifying 0.7), T equal to the temperature in Kelvin or Rankin, and A and B being constants for a given oil, then

$$\log_{10}(\log_{10} Z) = A + B(\log_{10} T) \tag{2.44}$$

Sanchez-Rubio et al. [57] have suggested an alternative method in which the Walther equation is used. In this case, they define a viscosity number (VN) as follows:

$$\mathrm{VN} = \left[1 + \frac{(3.55 + B)}{3.55}\right] \times 100 \tag{2.45}$$

The value of 3.55 was selected because lubricating oils with a VI of 100 have a value of B about equal to -3.55. Using this expression implies that VN = 200 would correspond to an idealized oil whose viscosity has no dependence of viscosity on temperature (i.e., $B = 0$). The pressure to which an oil is subjected to

Table 2.12 Effects of temperature and pressure on viscosity of selected lubricants having various viscosity indexes (All fluids have viscosities of 20 °C St at 40 °C and 0.1 MPa Pressure) [31] (Reproduced with permission)

Quantity	Fluorolube	Hydrocarbon	Ester	Silicone
Viscosity index	−132	100	151	195
Viscosity (cSt) at −40 °C	500,000	14,000	3,600	150
Viscosity (cSt) at −100 °C	2.9	3.9	4.4	9.5
Viscosity (cSt) at −40 °C and 138 MPa	2700	340	110	160
Viscosity (cSt) at −40 °C and 552 MPa	>1,000,000	270,000	4,900	48,000

can influence its viscosity. The relationship between dynamic viscosity and hydrostatic pressure p can be represented by

$$\eta = \eta_o \exp(\alpha p) \tag{2.46}$$

where η and α vary with the type of oil. Table 5.12 illustrates the wide range of viscosities possible for several liquid lubricants under various temperatures and pressures. The viscosity indices for these oils range from −132 to 195. Viscosity has a large effect on determining the regime of lubrication and the resultant friction coefficient. Similarly to the effect of strain rate on the shear strength of certain metals, like aluminium, the rate of shear in the fluid can also alter the viscosity of a lubricant.

Ramesh and Clifton [53] constructed a plate impact device to study the shear strength of lubricants at strain rates as high as 900,000/sec and found significant effects of shear rate on the critical shear stress of lubricants. In a Newtonian fluid, the ratio of shear stress to shear strain does not vary with stress, but there are other cases, such as for greases and solid dispersions in liquids, where the viscosity varies with the rate of shear. Such fluids are termed non-Newtonian and the standard methods for measuring viscosity cannot be used.

Lubrication regimes determine the effectiveness of fluid film formation, and hence, surface separation. In the first decade of the twentieth century, Stribeck developed a systematic method to understand and depict regimes of journal bearing lubrication, linking the properties of lubricant viscosity (η), rotational velocity of a journal (ω), and contact pressure (p) with the coefficient of friction. Based on the work of Mersey, McKee, and others, the dimension-less group of parameters has evolved into the more recent notation (*ZN/p*), where Z is viscosity, *N* is rotational speed, and *p* is pressure. The Stribeck curve has been widely used in the design of bearings and to explain various types of behavior in the field of lubrication. At high pressures, or when the lubricant viscosity and/or speed are very low, surfaces may touch, leading to high friction. In that case, friction coefficients are typically in the range of 0.5–2.0. The level plateau at the left of the curve represents the boundary lubrication regime in which friction is lower than for unlubricated sliding contact ($\mu = 0.05$ to about 0.15). The drop-off in friction is called the mixed film regime. The mixed regime refers to a combination of boundary lubrication with hydrodynamic or elastohydrodynamic lubrication.

Beyond the minimum in the curve, hydrodynamic and elastohydrodynamic lubrication regimes are said to occur. Friction coefficients under such conditions can be very low. Typical friction coefficients for various types of rolling element bearings range between 0.001 and 0.0018.

The conditions under which a journal bearing of length L, diameter D, and radial clearance C (bore radius minus bearing shaft radius) operates in the hydrodynamic regime can be summarized using a dimensionless parameter known as the Sommerfeld number S, defined by

$$S = \frac{\eta NLD}{P}\left(\frac{R}{C}\right)^2 \tag{2.47}$$

where P is the load on the bearing perpendicular to the axis of rotation, N is the rotational speed, η is the dynamic viscosity of the lubricant, and R is the radius of the bore. The more concentrically the bearing operates, the higher the value of S, but as S approaches 0, the lubrication may fail, leading to high friction. Sometimes Stribeck curves are plotted using S instead of (ZN/p) as the abscissa. Raimondi and his co-workers [52] added leakage considerations when they developed design charts in which the logarithm of the Sommerfeld number is plotted against the logarithm of either the friction coefficient or the dimensionless film thickness.

Using small journal bearings, McKee developed the following expression for the coefficient of friction μ based on the journal diameter D, the diametral clearance C, and an experimental variable k, which varies with the length to diameter ratio (L/D) of the bearing (Hall et al. [24]:

$$\mu = \left(4.73 \times 10^{-8}\right)\left(\frac{ZN}{p}\right)\left(\frac{D}{C}\right) + k \tag{2.48}$$

The value of k is about 0.015 at $(L/D) = 0.2$, drops rapidly to a minimum of about 0.0013 at $(L/D) = 1.0$, and rises nearly linearly to about 0.0035 at $(L/D) = 3.0$. A simpler expression, discussed by Hutchings [28], can be used for bearings that have no significant eccentricity:

$$\mu = \frac{2\pi}{S}\frac{h}{R} \tag{2.49}$$

where S is the Sommerfeld number, h is the mean film thickness, and R is the journal radius. With good hydrodynamic lubrication and good bearing design, μ can be as low as 0.001.

Hydrodynamic lubrication, sometimes called thick-film lubrication, generally depends on the development of a converging wedge of lubricant in the inlet of the interface. This wedge generates a pressure profile to force the surfaces apart. When the elastic deformation of the solid bodies is similar in extent to the thickness of the lubricant film, then elastohydrodynamic lubrication is said to occur. This latter regime is common in rolling element bearings and gears where high Hertz contact stresses occur. If the contact pressure exceeds the elastic limit of the surfaces, plastic deformation and increasing friction occur. One way to understand and control the various lubrication regimes is by using the specific film thickness (also called the lambda ratio), defined as the ratio of the minimum film thickness in the interface (h) to the composite root-mean-square (rms) surface roughness σ^*:

$$\wedge = h/\sigma* \tag{2.50}$$

where the composite surface roughness is defined in terms of the rms roughness ($\sigma_{1,2}$) of surfaces 1 and 2, respectively:

$$\sigma* = \sqrt{\left(\sigma_1^2 + \sigma_2^2\right)} \tag{2.51}$$

For the boundary regime, $\Lambda \ll 1$. For the mixed regime $1 < \Lambda < 3$. For the hydrodynamic regime, $\Lambda \gg 6$, and for the elastohydrodynamic regime, $3 < \Lambda < 10$. Boundary lubrication produces friction coefficients that are lower than those for unlubricated sliding but higher than those for effective hydrodynamic lubrication, typically in the range $0.05 < \mu < 0.2$. Briscoe and Stolarski [10] have reviewed friction under boundary-lubricated conditions. They cited the earlier work of Bowden, which gave the following expression for the friction coefficient under conditions of boundary lubrication:

$$\mu = \beta\mu_a + (1 - \beta)\mu_1 \tag{2.52}$$

where the adhesive component μ_a and the viscous component of friction μ_1 are given in terms of the shear stress of the adhesive junctions in the solid (metal) τ_m and the shear strength of the boundary film τ_1 under the influence of a contact pressure σ_p:

$$\mu_a = \frac{\tau_m}{\sigma_p} \tag{2.53}$$

The parameter, β, is called the fractional film defect [10] and is:

$$\beta = 1 - \exp\left\{-\left[\frac{(30.9 \times 10^5)T_m^{1/2}}{VM^{1/2}}\right]\exp\left(-\frac{E_c}{RT}\right)\right\} \tag{2.54}$$

where M is the molecular weight of the lubricant, V is the sliding velocity, T_m is the melting temperature of the lubricant, E_c is the energy to desorb the lubricant molecules, R is the universal gas constant, and T is the absolute temperature. Various graphical methods have been developed to help select boundary lubricants and to help simplify the task of bearing designers. Most of these methods are based on the design parameters of bearing stress (or normal load) and velocity. One method, developed by Glaeser and Dufrane [23] involves the use of design charts for different bearing materials. An alternate but similar approach was used in developing the so-called IRG transitions diagrams (subsequently abbreviated ITDs), an approach that evolved in the early 1980s, was applied to various bearing steels, and is still being used to define the conditions under which boundary-lubricated tribosystems operate effectively. Instead of pressure, load is plotted on the ordinate. Three regions of ITDs are defined in terms of their frictional behavior: Region I, in which the friction trace is relatively low and smooth, Region II, in which the friction trace begins with a high level then settles down to a lower, smoother level, and Region III, in which the friction trace is irregular and remains high. The transitions between Regions I and II or between Regions I and III are described as a collapse of liquid film lubrication. The locations of these transition boundaries for steels were seen to depend more on the surface roughness of the materials and the composition of the lubricants and less on microstructure and composition of the alloys. Any of the following testing geometries can be used to develop ITDs: four-ball machines, ball-on-cylinder machines, crossed-cylinders machines, and flat-on-flat testing machines (including flat-ended pin-on-disk). One important aspect of the use of liquid lubricants is how they are applied, filtered, circulated, and replenished. Lubricants can also be formed on surfaces by the chemical reaction of vapor-phase precursor species in argon and nitrogen environments.

Acknowledgments The authors acknowledge with thanks the permission to reproduce Sects. 2 and 3 from Woodhead Publishing, Cambridge, UK. The sections were adapted from the publication: M. J. Jackson and J. S. Morrell, 'Tribology in Manufacturing', Chapter 5, p.p. 161-242, published in 'Tribology for Engineers: A Practical Guide', Edited by J.Davim, Woodhead Publishing, Cambridge, UK, 2011. *ISBN 0 85709 114X and ISBN-13: 978 0 85709 114 7.*

References

1. Ackroyd B (1999) Ph.d. Thesis, School of Industrial Engineering, Purdue University, West Lafayette, Indiana
2. Akhmatov AS (1939) Some items in the investigation of the external friction of solids, Trudy Stankina; cited by I. V. Kragelski (1965) in Friction and Wear. Butterworths, London, p 159
3. Baglin JEE, Gottschall RJ, Batich CD (1987) Materials Research Society, Pittsburgh, pp 81–86
4. Bednar MS, Cai BC, Kuhlmann-Wilsdorf D (1993) Pressure and structure dependence of solid lubrication. Lub. Eng. 49(10):741–749
5. Bisson EE, Anderson WJ (1964) Advanced bearing technology, NASA SP-38. National Aeronautics and Space Administration, Washington
6. Black PH (1961) Theory of Metal Cutting. McGraw-Hill, New York, pp 45–72
7. Blok H (1963) The flash temperature concept. Wear 6:483–494
8. Boehringer RH (1992) Grease, in ASM handbook, friction, lubrication, and wear technology, ASM International, Materials Park Ohio, 10th edn, vol 18, pp 123–131
9. Bowden FP, Tabor D (1986) The Friction and Lubrication of Solids. Clarendon Press, Oxford
10. Briscoe BJ (1992) Friction of organic polymers. In: Singer IL, Pollock HM (eds) Fundamentals of friction: macroscopic and microscopic processes. Kluwer, Dordrecht, pp 167–182
11. Briscoe BJ, Stolarski TA (1993) Friction, chapter 3 in characterization of tribological materials. In: Glaeser WA, Butterworth Heinemann, Boston, pp 48–51
12. Buckley DF (1981) Surface effects in adhesion friction, wear, and lubrication,chapter5. Elsevier, New York, pp 245–313
13. Burton RA (1980) Thermal deformation in frictionally-heated systems. Elsevier, Lausanne, p 290
14. Campbell WE (1940) Remarks printed in proceedings of MIT conference on friction and surface finish. MIT Press, Cambridge, p 197

15. Cowan RS, Winer WO (1992). Frictional heating calculations, in ASM Hand-book. Friction, lubrication, and wear technology, ASM International, Materials Park, Ohio, 10th edn, vol 18, pp 39–44
16. Dow TA, Stockwell RD (1977) Experimental verification of thermoelastic instabilities in sliding contact. J Lubric Technol 99(3):359
17. Doyle ED, Horne JG, Tabor D (1979) Frictional interactions between chip and rake face in continuous chip formation. Proc Roy Soc London A 366:173–187
18. Enthoven JC, Cann PM, Spikes HA (1993) Temperature and scuffing. Tribol Trans 36(2):258–266
19. Ferrante J, Bozzolo GH, Finley CW, Banerjea A (1988). Interfacial adhesion: theory and experiment. In: Mattox DM, Baglin JEE, Gottschall RJ, Batich CD (eds) Adhesion in solids, Materials Research Society, Pittsburgh, pp 3–16
20. Greenwood JA, Williamson JBP (1966) Contact of nominally flat surfaces. Proc Roy Soc London A 295:300–319
21. Greenwood JA (1992) Problems with surface roughness. In: Singer IL,. Pollock HM, Fundamentals of friction: macroscopic and microscopic processes, Kluwer, Dordrecht, pp 57–76
22. Griffioen JA, Bair S, Winer WO (1985). Infrared surface temperature in a sliding ceramic-ceramic contact. In: Dowson D et al. (eds)Mechanisms of surface distress, Butterworths, London, pp 238–245
23. Glaeser WA, Dufrane KF (1978) New design methods for boundary lubricated sleeve bearings. Machine design, pp 207–213
24. Hall AS, Holowenko AR, Laughlin HG (1961) Lubrication and bearing design, in machine design schaum's outline series. McGraw-Hill, New York, p 279
25. Hirst W, Hollander AE (1974) Surface finish and damage in sliding. Proc Roy Soc London A 233:379
26. Hokkirigawa K, Kato K (1988) An experimental and theoretical investigation of ploughing, cutting and wedge formation during abrasive wear. Tribol Int 21(1):51–57
27. Horne JG, Doyle ED, Tabor D (1978) Direct observation of the contact and lubrication at the chip-tool interface. In: Proceedings of the first international conference on lubrication challenges in metalworking and processing, IIT Research Institute, Chicago
28. Hutchings IM (1992) Tribology—friction and wear of engineering materials. CRC Press, Boca Raton, p 65
29. Hwang J (2005) Ph.d. thesis, School of Industrial Engineering, Purdue University, West Lafayette, Indiana
30. Israelachvili JN (1992)\ Adhesion, friction, and lubrication of molecularly smooth surfaces. In: Pollock HM, Singer IL (eds) Fundamentals of friction: macroscopic and microscopic processes, Kluwer, Dordrecht, pp 351–381
31. Jackson MJ, Morrell JS (2011) Tribology in manufacturing, chapter 5. In: Davim J (ed) Tribology for engineers: a practical guide, Woodhead Publishing, Cambridge, pp 161–242
32. Jaeger JC (1942) Moving sources of heat and the temperature at sliding contacts. J. Proc. Royal Soc. N. South Wales, 76:203–224
33. Klaus EE, Tewksbury EJ (1984) Liquid lubricants. In: Booser ER (ed) The handbook of lubrication, theory and practice of tribology, vol II, CRC Press, Boca Raton, pp 229–254
34. Kosterin JI, Kraghelski IV (1962) Rheological phenomena in dry friction. Wear 5:190–197
35. Kragelski IV (1965) Friction and wear. Butterworths, London, p 200
36. Kragelskii IV, Dobychin MN, Kombalov VS (1982) Friction and wear calculation methods. Pergamon Press, Oxford, pp 178–180
37. Kudinov VA, Tolstoy DM (1986). Friction and oscillations. In: Kragelski (ed)Tribology handbook, Mir, Moscow, p 122
38. Kuhlmann-Wilsdorf D (1987) Demystifying flash temperatures I. analytical expressions based on a simple model. Mater Sci Eng 93:107–117
39. Landman U, Luetke WD, Burnham NA, Colton RJ (1990) Atomistic mechanisms and dynamics of adhesion, nanoindentation, and fracture. Science 248:454–461

40. Lee S (2006), Ph.d. thesis, School of Industrial Engineering, Purdue University, West Lafayette, Indiana
41. McClelland GM, Mate CM, Erlandsson R, Chiang S (1988). Direct observation of friction at the atomic scale. In: Mattox DM, Baglin JEE, Gottschall RJ, Batich CD (eds) Adhesion in solids, Materials Research Society, Pittsburgh, pp 81–86
42. McClelland GM, Mate CM, Erlandsson R, Chiang S (1987) Atomic-scale friction of a tungsten tip on a graphite surface. Phys Rev Lett 59:1942
43. McCool J (1986) Comparison of models for the contact of rough surfaces. Wear 107:37–60
44. Madhavan V (1993) M.S. Thesis, School of Industrial Engineering, Purdue University, West Lafayette, Indiana
45. Miyoshi K, Buckley DH (1981) Anisotropic tribological properties of silicon carbide. In: Proceedings of wear of the materials, ASME, New York, pp 502–509
46. Moore DF (1975) Principles and applications of tribology. Pergamon Press, Oxford, p 152
47. Mulhearn TO, Samuels LE (1962) Wear 5:478
48. Pollock HM, Singer IL (eds) (1992) Fundamentals of friction: macroscopic and microscopic processes, Kluwer, Dordrecht, pp 621
49. Quinn TFJ, Winer WO (1987) An experimental study of the 'hot spots' occurring during the oxidational wear of tool steel on sapphire. J Tribol 109(2):315–320
50. Rabinowicz E (1965) Friction and wear of materials. John Wiley and Sons, New York, p 69
51. Rabinowicz E (1992) Friction coefficients of noble metals over a range of loads. Wear 159:89–94
52. Raimondi AA (1968) Analysis and design of sliding bearings, Chapter 5 in Standard Handbook of Lubrication Engineering. McGraw-Hill, New York
53. Ramesh KT, Clifton RJ (1987) A pressure-shear plate impact experiment for studying the rheology of lubricants at high pressures and high shear rates. J. Tribal. 109:215–222
54. Robbins MO, Thompson PA (1991) The critical velocity of stick-slip motion. Science 253:916
55. Robbins MO, Thompson PA, Grest GS (1993) Simulations of nanometer-thick lubricating films Mater Res Soc Bull XVIII(5):45–49
56. Robinson GM (2007) Ph. d. thesis, College of Technology, Purdue University, West Lafayette, Indiana
57. Sanchez-Rubio M, Heredia-Veloz A, Puig JE, Gonzalez-Lozano S (1992) A better viscosity-temperature relationship for petroleum products. Lubr Eng 48(10):821–826
58. Sikorski ME (1964) The adhesion of metals and factors that influence it. In: Bryant PJ, Lavik L, Salomon G (eds) Mechanisms of solid friction (eds) Elsevier, Amsterdam, pp 144–162
59. Song JF, Vorburger TV (1992) Surface texture, in ASM handbook,friction, lubrication, and wear technology, ASM International, Materials Park, Ohia, 10th edn, vol 18, pp 334–345
60. Suh NP (1986) Tribophysics. Prentice-Hall, Englewood Cliffs, pp 416–424
61. Thompson PA, Robbins MO (1990) Origin of stick-slip motion in boundary lubrication. Science 250:792–794
62. Wills JG (1980) Lubrication fundamentals. Marcel Dekker, New York
63. Wheeler DR (1975) The effect of adsorbed chlorine and oxygen on the shear strength of iron and copper junctions, NASA TN D-7894
64. Whitehouse DJ, Archard JF (1970) The properties of random surfaces of significance in their contact. Proc Royal Soc London A 316:97–121
65. Young WC (1989) Roark's formulas for stress and strain, 6th edn. McGraw-Hill, New York
66. Bhushan B, Gupta BK (1991) Handbook of tribology, McGraw Hill
67. Handbook of chemistry and physics(1967). CRC Press, 48th edn, p 3
68. Rabinowicz E (1971) Plate sliding on inclined plate at 50% rel. humidity ASLE Trans., vol 14, p 198
69. Friction data guide general magnaplate corp (1988) Ventura, California 93003, TMI model 98-5 slip and friction tester, 200 grams load, ground specimens, 54 % rel. humid. average of 5 tests

Chapter 3
Tribology in Metal Forming Processes

Sergio Tonini Button

Abstract This chapter presents the main aspects of the tribological systems observed in the most important metal forming processes. Friction, wear and lubrication are discussed in general terms to describe friction models, wear mechanisms and lubrication regimes commonly found in metal forming, which are analyzed for each of the five groups of processes (rolling, extrusion, forging, drawing and stamping). Recent developments concerning to environmental and cost reduction are also discussed.

3.1 Introduction

Tribology plays a key role in the design of systems which represent metal forming processes due to its influence on many parameters which define the products quality, equipment capacity, tools life, environment sanity and workers health.

The interrelation of the several parameters involved in a process, related to raw material, tools, equipment, and products is directly influenced by tribological aspects mainly represented by the friction in the interface workpiece-tools, dictated by the lubrication regime established in the interface.

A lot of industrial and academic research has been done to evaluate the tribological conditions, and to propose new lubrication methods, and lubricants, to minimize the deleterious effects of friction on forming loads, tools life and products quality.

S. T. Button (✉)
Department of Materials Engineering, School of Mechanical Engineering
University of Campinas, Campinas, Brazil
e-mail: sergio1@fem.unicamp.br

J. P. Davim (ed.), *Tribology in Manufacturing Technology*, Materials Forming, Machining and Tribology, DOI: 10.1007/978-3-642-31683-8_3,

To understand how tribology influences metal forming two concepts have to be analyzed: what is friction and how lubrication can lead to sound processes.

3.2 Friction

Most of metal forming processes present at least one moving tool in respect to the workpiece surface. The contact pressure in the interface tool-workpiece is always very high in order to plasticize the workpiece material and consequently define the product shape, often with intricate details and significant change of overall dimensions.

The effects of the high contact pressure can be enhanced by the topography of the surfaces, especially of the workpiece, and by the high local temperatures usually found in metal forming.

The influence of the friction on the tools wear and products surface quality is more pronounced in processes like hot forging and hot extrusion which present a difficult entrapment of the lubricant into the deformation zone.

Otherwise, processes like wire drawing and sheet stamping present low friction coefficients because of the light deformations, relatively low temperatures and a plenty and available source of lubricant constantly dragged and entrapped between tools and workpiece surface.

In some processes like sheet hot rolling and cross-wedge-rolling friction is necessary despite all the consequent problems it causes. For instance, rolling is not possible if the friction coefficient is less than a minimum value, as it can be seen in Fig. 3.1: high coefficients cause an excessive increase of the specific pressure, while a small coefficient tends to move the neutral section, where occurs the transition from backward to forward slip, to the exit of the arc of contact, and therefore the plate tends to backward slip without rolling [1].

Direct measurement of the friction along the process is practically impossible no matter what process is being analyzed. The most common technique widely applied is to measure forming loads and then calculate the friction coefficient with expressions obtained by analytical and empirical methods. But this method does not correctly represent processes with severe conditions, hot forging and hot extrusion for instance, and laboratory experiments must be held to estimate the friction coefficient.

Many experimental methods and apparatus are reported in the literature with this objective [2–4]. They are designed to replace the widely used and limited pin-on-disk (or ball-on-disk) [5] and draw-bend friction tests [6], as well to best represent many processing conditions: tools and workpiece materials, temperature, surface roughness, relative velocity, lubricant and lubrication regime, and contact pressure. These methods also provide indirect measurements and again, the friction coefficient is an estimative.

Two models are usually applied to define the coefficient in metal forming. The first is the well-known principle first observed by Leonardo da Vinci (c.a. 1500)

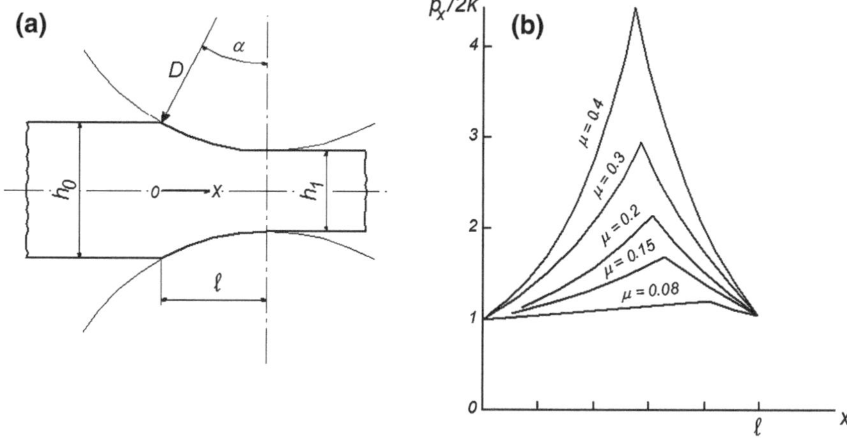

Fig. 3.1 **a** Plate rolling **b** Specific pressure distribution along the axis in the contact length l as a function of friction coefficient (μ) for $h_1/h_0 = 1.3$ and $h_1/D = 0.012$

and formulated in the Amontons-Coulomb law resulted from the works of Guillaume Amontons (c.a. 1700) and Charles August Coulomb (c.a. 1785) that states: "the resistance to the relative movement due to friction is proportional to the normal compressive force applied to the surfaces". As observed by da Vinci, the areas in contact have no effect on friction, and then the friction coefficient (μ) can be calculated in terms of tangential (τ) and normal (σ) stresses as:

$$\mu = \tau/\sigma \tag{3.1}$$

There are many references in the literature with tables of friction coefficient based in the Amontons-Coulomb law for many materials combinations and geometries, but all the tables were obtained under controlled experimental conditions, and most important, with relatively low contact pressures.

They are valid only for processes which present small deformations, low temperatures and good lubrication conditions, as observed in wire drawing and sheet forming: tool and workpiece present small surface roughness, and the supply of lubricant is continuous and effective. Under these conditions, low contact pressures are established and the Amontons-Coulomb law can be applied with reasonable significance.

Otherwise, in processes with opposite conditions, i.e. high temperatures, high surface roughness and difficult access of lubricant to the deformation zone, very high contact pressures are present, and in many processes like hot forging and hot extrusion, the workpiece material flows sub-superficially under high shear stresses.

In those conditions the Amontons-Coulomb law has no physical meaning since the tangential stress (τ) reaches a maximum and constant value equal to the yield stress in pure shear (k). Expression (3.1) may be rewritten as

$$\mu = k/\sigma \tag{3.2}$$

As the normal stress increases during the process, the tangential stress remains constant, and then an inconsistent reduction of the friction coefficient is obtained. To better represent these severe conditions a second model was proposed with a constant friction factor (m) defined from the von Mises flow criterion, as follows:

$$m = \tau/k \tag{3.3}$$

where $k = 0.577\ \sigma_0$ and σ_0 is the flow stress of the workpiece material under tension or compression, and m can be found from zero, for perfectly lubricated interfaces, to the unity when a complete adhesion tools-workpiece is established.

Considering these limits and substituting expression (3.3) into expression (3.2), the friction coefficient (μ) can be found between zero (no friction—free sliding) and 0.577 (maximum friction—complete adhesion).

Well-lubricated processes like wire drawing and deep drawing present friction coefficients between 0.05 and 0.15 while severe processes like hot rolling and hot forging may present coefficients up to 0.4. In some processes which adhesion is very common, like in hot extrusion of steel bars, constant factor (m) is more representative than friction coefficient (μ).

Bowden and Tabor (c.a. 1950) demonstrated that the true area of contact, represented by the asperities of both surfaces, is a very small percentage of the apparent contact area, and that tangential stresses caused by friction are directly proportional to the true contact area [7].

Their work initiated the researches on micro-scale mechanisms to describe tribological systems based on the interaction of asperities and lubricants which influences the lubrication regimes present in metal forming processes.

Under some lubrication conditions a full hydrodynamic regime can be established and the interface is fulfilled with a continuous lubricant film, as observed in the high speed drawing of steel fine wires. In this case, the hypothesis behind the Amontons-Coulomb law that kinetic friction is not affected by the sliding velocity (U) is not valid since the frictional sliding force (F) is dependent on that velocity as shown in the Reynold′s steady state equation of fluid film lubrication:

$$F \propto (U\eta/h) \tag{3.4}$$

Where η is the lubricant dynamic viscosity and h is the film thickness.

Therefore, in hydrodynamic lubrication the friction coefficient must be calculated by the curve proposed by Richard Stribeck (c.a. 1900) shown in Fig. 3.2, where friction coefficient (μ) is a function of S (Stribeck's number) defined by the surfaces relative velocity (U), the lubricant viscosity (η) and by the interface load (F) for the three lubrication regimes: hydrodynamic (thick and thin film), mixed and boundary lubrication [8].

With low velocities (region 1), few lubricant is carried to the interface and the boundary lubrication is established, direct metallic contact can occur and high friction coefficients are then observed. With the increase of velocity, the film pressure is raised, more lubricant is carried to the interface, filling the valleys,

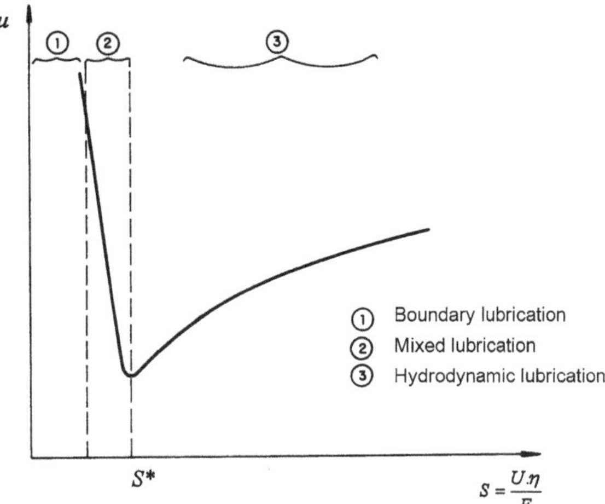

Fig. 3.2 Stribeck's diagram

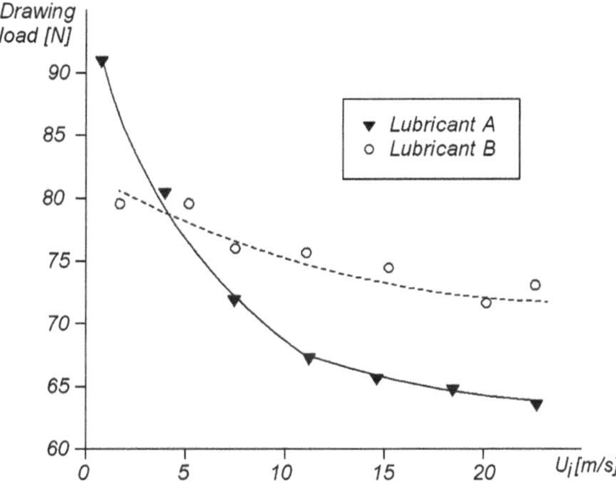

Fig. 3.3 Wire drawing load versus initial drawing velocity for two lubricants: A-$\eta_0 = 1.001$ Ns/m2, $\alpha = 2.0.10^{-8}$ m^2/N; B-$\eta_0 = 0.011$ Ns/m^2, $\alpha = 0.7.10^{-8}$ m^2/N

promoting a mixed lubrication regime (region 2), with an important reduction of friction.

Above a critical value of S, defined as S*, the roughness peaks are flattened, more and more lubricant is carried to the interface and thus the hydrodynamic regime is established, the surfaces are separated and the friction coefficient reduced.

In the region 3, the increase of friction coefficient with velocity is due to large amounts of lubricant which are carried to the interface increasing the pressure, the film thickness and the lubricant viscosity, which consequently increase of the sliding force (F).

Figure 3.3 shows experimental load results as a function of the sliding velocity and lubricant viscosity in the wire drawing of annealed stainless steel AISI 304L with 19 % of cross section reduction and die angle (2β) equal to 10°.

At low velocities both lubricants present high loads, and the low viscosity lubricant presents the smaller load because it is easier to be dragged to the deformation zone even at very slow conditions.

Confirming the Stribeck's curve, both lubricants present a reduction of the drawing load with the increase of sliding velocity, which is more significant for the high viscosity lubricant, because its capacity to build up the lubricant pressure and to establish a thicker and continuous film.

3.3 Lubrication

Like in other forming processes and industrial applications, lubricants are used in metal forming to keep apart two surfaces with relative motion in order to minimize friction, and consequently the heat generation and the wear of the surfaces in contact.

The correct choice of the lubricant and lubrication conditions depends on several factors such as processing temperature and velocity, physical and chemical compatibility of the lubricants and tools and workpiece materials, and on how easy is to apply and to remove the lubricant, e.g., for some stamped products used in the food industry it is not allowed the presence of contaminants common in industrial lubricants, and in this case fine polymers coatings are used instead fluid lubricants [9, 10].

More recently, environmental concerns were included in the selection of metal forming lubricants to avoid elements like phosphorus, sulphur and chlorine usually found as additives in extreme pressure lubricants [11–13], and zinc widely used in phosphate conversion coating [14–16].

Four groups of lubricants are used in metal forming processes:

- fluids with a large viscosity range, from the light mineral oils to synthetic greases, which form a continuous film and promote a hydrodynamic lubrication;
- emulsions with appropriate compositions which reacts chemically with the tools and workpiece generating very thin films and establishing a boundary lubrication
- organic materials with additives for extreme pressures which reacts with the tools and workpiece to form compound layers at the contact interface;
- solids like graphite, glass, boron nitride and molybdenum disulfide, used to separate the tools and workpiece surfaces.

Figure 3.4 shows four different lubrication regimes which may be present at the tool-workpiece interface. These regimes depend on process variables like tool and workpiece surface roughness, lubricant viscosity and compressibility, temperature, velocities and deformation [17].

In hydrodynamic lubrication the surfaces are completely separated by a lubricant film (Fig. 3.4a) many times thicker than the roughness of both surfaces, and than the molecular size of the lubricant.

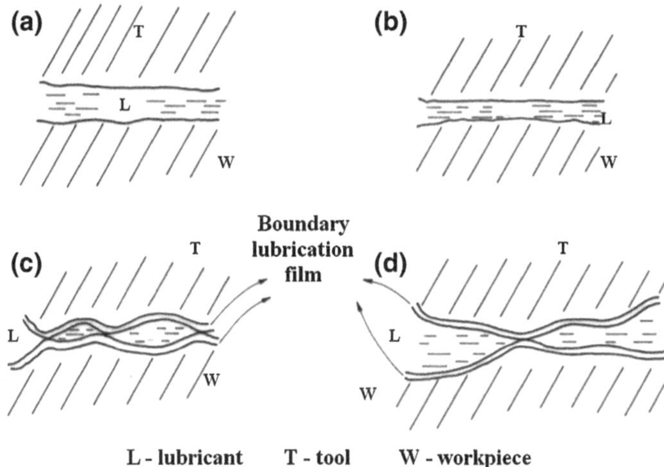

Fig. 3.4 Lubrication regimes in metal forming

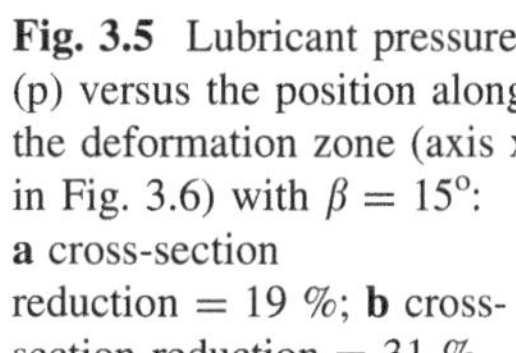
Fig. 3.5 Lubricant pressure (p) versus the position along the deformation zone (axis x in Fig. 3.6) with $\beta = 15^{o}$: **a** cross-section reduction = 19 %; **b** cross-section reduction = 31 %

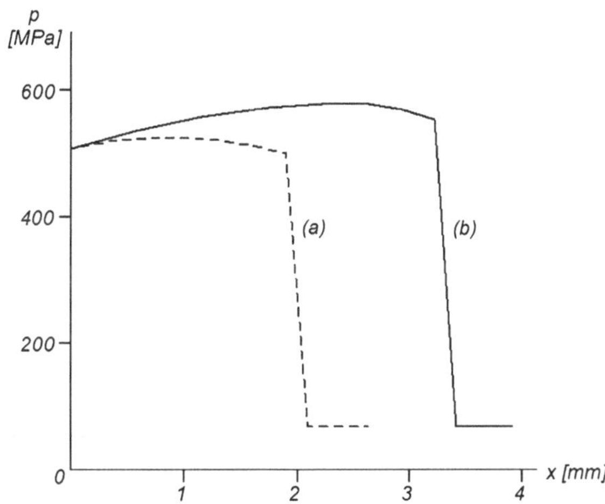

In this regime a uniform and continuous lubricant film is formed when the pressure is built up at the entrance of the deformation region as shown in Fig. 3.5 for the cold extrusion of a AISI 1020 steel with a medium pressure mineral oil: the pressure in the lubricant is high for the greater area reduction as well the pressure built up occurs early near the deformation zone entrance.

The resistance to the relative movement is defined by the chemical–mechanical properties of the lubricant. Friction coefficients lower than 0.05 are observed, and the wear of the tools is reduced.

Otherwise some problems may arise due the formation of a thick film. Hard debris can be dragged to the interface reducing the lubricant efficiency and damaging the workpiece surface.

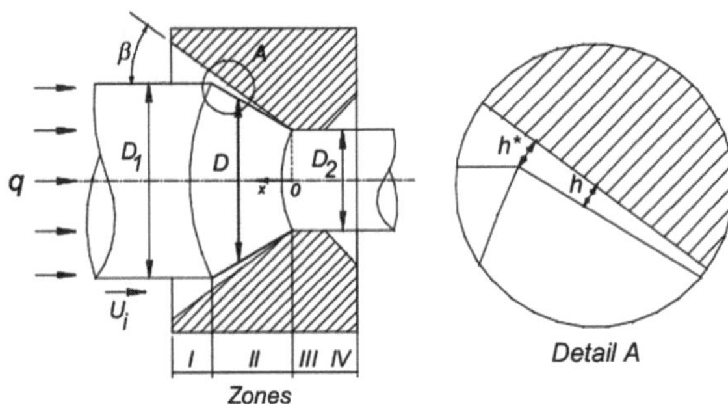

Fig. 3.6 Regions of the extrusion die

If the film thickness is increased with the viscosity, the asperities of the workpiece surface will be not flattened by the tools surface and therefore being free to deform will increase the product surface roughness [18], as observed in rolling of aluminum strips [19].

With thinner lubricant films (Fig. 3.4b)–thickness between three and ten times the mean average roughness of the surfaces–the hydrodynamic lubrication also separates the surfaces but some peaks may be in contact and therefore rough surfaces can break down the film and establish a direct metallic contact, increasing forming loads and tool wear.

In both previous regimes, the film thickness (h^*) at the entrance of the deformation region (Zone II in Fig. 3.6) is strongly influenced by many processing parameters, and it is increased when the lubricant viscosity, the sliding velocity, and the yield strength of the workpiece are increased [20], as defined by Eq. (3.5) for cold extrusion, and shown in Fig. 3.7: in the wire drawing of annealed stainless steel AISI 304 L with a cross section reduction of 19 % and with a high pressure mineral oil, the film thickness increases significantly with the reduction of the semi-angle of the die (β) and with the increase of the initial speed (U_i) because more lubricant is dragged and entrapped into the deformation zone [21].

$$h^* = \frac{3U_1\eta_0\alpha}{e^{-\alpha q}\tan\beta\ (1 - e^{-\alpha\sigma_0})} \tag{3.5}$$

where $q = 2\sigma_0 \ln\left(\frac{D_1}{D_2}\right)$ and η_0 is the initial viscosity and α is the pressure coefficient in the expression $\eta = \eta_0 e^{\alpha p}$

Thicker films may develop pressure instabilities causing the formation of "lubricant pockets" which deform locally the workpiece surface worsening it and increasing surface roughness [22–24].

Boundary lubrication shown in Fig. 3.4c is represented by lubricant films with thickness equal to some molecular size of the lubricant. This regime is strongly influenced by the topology roughness and chemical-physical properties of the surfaces involved.

Under ideal process conditions friction coefficients of 0.1 can be observed. If the film is broken down and the supply of new lubricant is difficult, the friction is increased and coefficients near 0.4 may be observed.

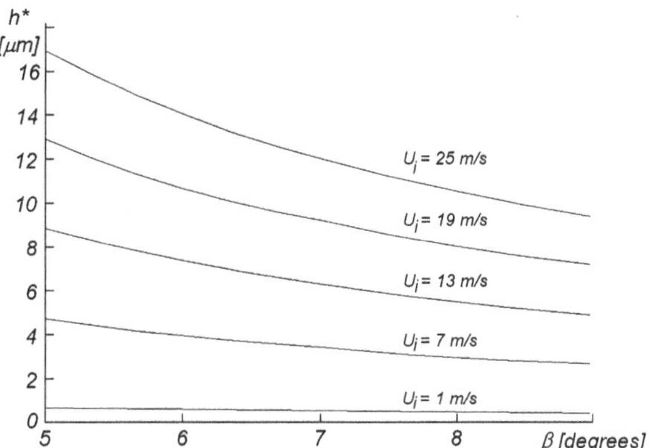

Fig. 3.7 Film thickness (h^*) versus die semi-angle (β) and initial drawing speed (U_i)

The mixed lubrication regime (Fig. 3.4d) is defined when the valleys between the roughness peaks are filled with sufficient lubricant that is continuously dragged to keep a stable and continuous lubricant film in the interface. Friction coefficients can vary from 0.05 (similar to hydrodynamic lubrication) to 0.4 if metallic contact occurs after film breakdown.

During sliding, lubricant is dragged out of the pockets and the pressure in the film is increased to a maximum value at the rear of the pockets and the lubricant tends to escape from the interface due to two mechanisms: micro-plasto-hydrodynamic lubrication (MPHDL) and micro-plasto-hydrostatic lubrication (MPHSL) [25], when the lubricants escapes in the dragging direction or in the opposite direction, respectively modifying the surface topography and roughness as shown in Fig. 3.8.

Both mechanisms depend on many factors previously discussed, like sliding velocity and lubricant viscosity, but also and perhaps most significantly, on the pocket geometry, which depends on the surface topography [26–28].

3.4 Wear

Every year metal forming industries lose a significant portion of their profits recovering and replacing damaged tools. The premature end of in-service tools life is often due to wear caused mainly by four mechanisms which work simultaneously: adhesion, abrasion, fatigue and chemical attack [29].

Adhesion occurs in processes which present high interface pressures and bad lubrication conditions, like those observed in hot forging and hot rolling, or in cold rolling of aluminum that is very reactive with the steel of the tools.

The direct contact of the sliding surfaces forms welded joints, and with the continuing deformation particles are ripped out from the tools surfaces.

These hard particles can be carried to the interface and cause *abrasion*, which can also be caused by deleterious debris like oxide particles present on the workpiece surface as observed in hot forging of steel components.

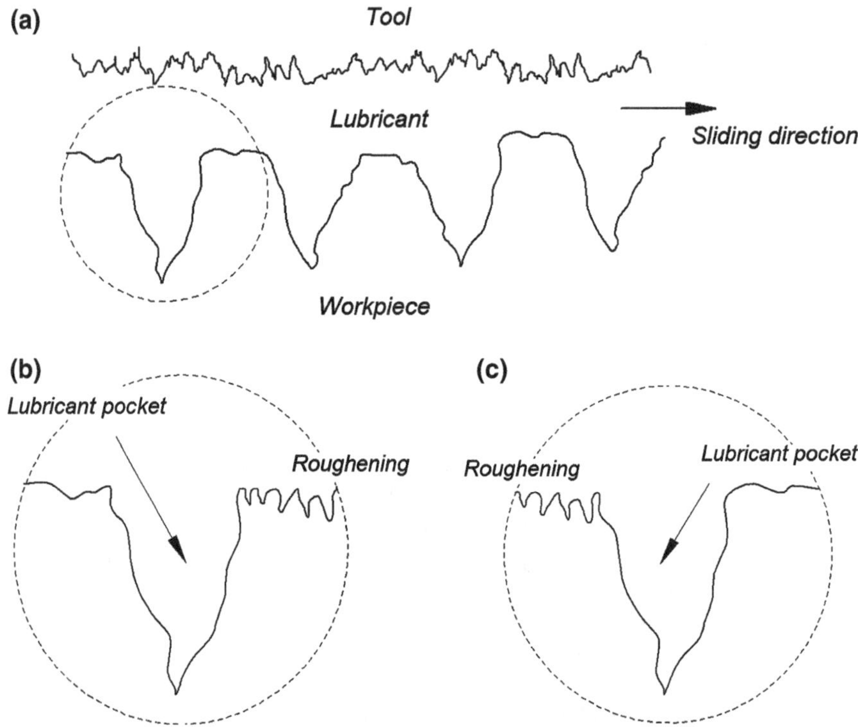

Fig. 3.8 Micro-lubrication mechanisms: **a** workpiece surface before sliding, **b** MPHDL, and **c** MPHSL

Most of forming processes present repeated cycles, like in hot forging of steel connecting rods. In each cycle the tools are submitted to high pressures during deformation, which are relieved when the forged part is ejected.

The temperature of tools surfaces is also dramatically changed during the cycles since the tools are heated during forging and instantly cooled before another workpiece is placed to be forged.

Figure 3.9 shows the gradients of contact pressure and von Mises equivalent stress in the forged connecting rod, and temperature within the forging dies.

The maximum contact pressure (Fig. 3.9a) and the maximum temperature (Fig. 3.9c) are observed at the center of the bottom end bearing at the end of each forging cycle. When the part is ejected and the dies are lubricated, the pressure is released and the temperature drops significantly. With successive strokes a cyclic variation of both contact pressure and temperature is established.

After hundreds or thousands of forging cycles, tools experience both mechanical and thermal *fatigue* [30], as shown in Fig. 3.10a [31].

The maximum equivalent stress is concentrated in the flash near to the end of the connecting region (Fig. 3.9b) caused by the significant restriction of material flow which causes the die to wear by abrasion and adhesion as shown in Fig. 3.10b.

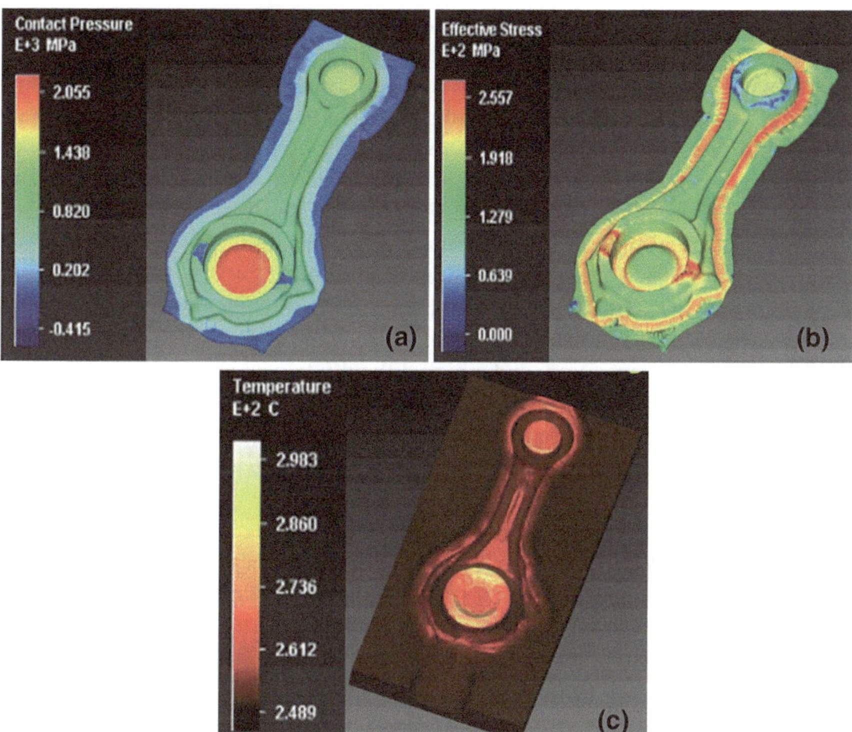

Fig. 3.9 Distribution of contact pressure (**a**) and von Mises equivalent stress (**b**) in the forged connecting rod, and temperature (**c**) in the lower die

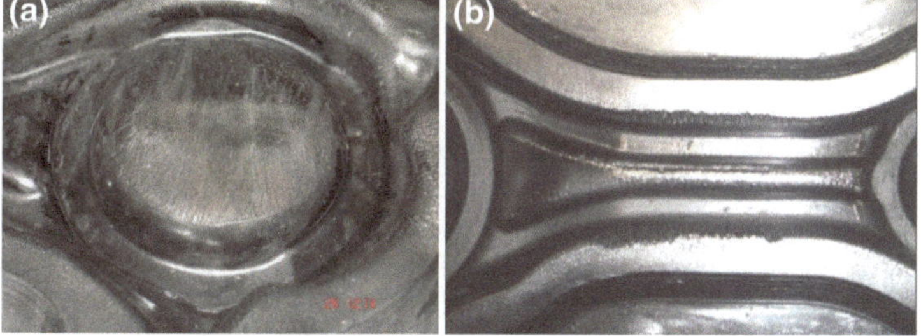

Fig. 3.10 Wear mechanisms observed in hot forging dies: **a** mechanical and thermal fatigue, and **b** abrasion and adhesion

Chemical attack occurs when the affinity of workpiece material with tools material and lubricant is increased by the contact pressure and the temperature at the sliding interface [32–34], then corrosion takes place and deleterious compounds are formed and carried to the interface increasing wear by adhesive and abrasive mechanisms [35]

To overcome the losses caused by tools wear and to increase tools in-service life hard coatings have been developed and applied to tools surface to reduce the affinity with lubricants and workpiece materials, and to promote a sound sliding avoiding adhesion, chemical attack and consequently abrasion. Some coatings are designed to present high toughness to reduce wear by fatigue.

Hard and wear-resistant coatings are obtained by CVD (Chemical Vapor deposition) of chromium nitrides and titanium nitrides, but the high deposition temperatures, usually near 1000 °C, make CVD inappropriate for coating tool steels which present temper temperatures below 500 °C. PVD (Physical Vapour Deposition) is an alternative method of hard coating with low deposition temperatures (below 500 °C) that present some disadvantages like high investments, a low rate of coating deposition, and a high difficulty to coat undercuts and intricate geometric details.

The deposition temperature can also be reduced by the superposition of plasma in the PACVD (Plasma-Assisted CVD), but the coating resistance is limited by the elasto-plastic behavior of the substrate.

Duplex treatments combining plasma diffusion and hard coatings, like plasma nitriding and PACVD (PN-PACVD), plasma nitriding and PVD (PN-PVD) and plasma nitriding with post-oxidizing (PN-PO), have been applied to forming tools with good resistance to wear, fatigue and corrosion, as well an improved compatibility to the substrate mechanical properties [36–40].

The surfaces of hot forging dies, made with H13 tool steel quenched and tempered, before and after one hundred strokes forging AISI 1020 steel at 1000 °C are shown in Fig. 3.11. One of the dies was coated by PN-PVD and the other, by PN-PO.

Before forging PN-PVD presents a uniform coating with some finely distributed white spots formed by the vapour deposition. The PN-PO shows an irregular distribution of compounds in the surface with white spots formed during nitriding and dark regions with oxide layers.

After forging the PN-PVD surface keeps the same uniformity observed before indicating a good wear resistance, while the PN-PO presented a heterogeneous surface with the coating compounds deformed and aligned to the material flow. The hardness numbers shown in Table 3.1 confirms the better performance of the PN-PVD coating that presents the best results before and after forging. The high standard deviation observed in the PN-PO coating before forging may be explained by heterogeneity observed in Fig. 3.11c.

More recently cladding techniques with carbon dioxide lasers were applied to incorporate titanium carbides and other hard particles into hot working tool steels to obtain micro-textured coatings with smooth surface, high strength and wear resistance, which present an excellent bonding of the coating layer to the substrate, better than the deposition techniques previously discussed [41, 42].

The main disadvantages of laser cladding are the high investments costs and the difficult control of the process parameters to obtain a uniform coating thickness [43].

Another technique recently adopted to enhance lubrication efficiency is surface texturing by machining, shot peening and laser to form pockets where the lubricant

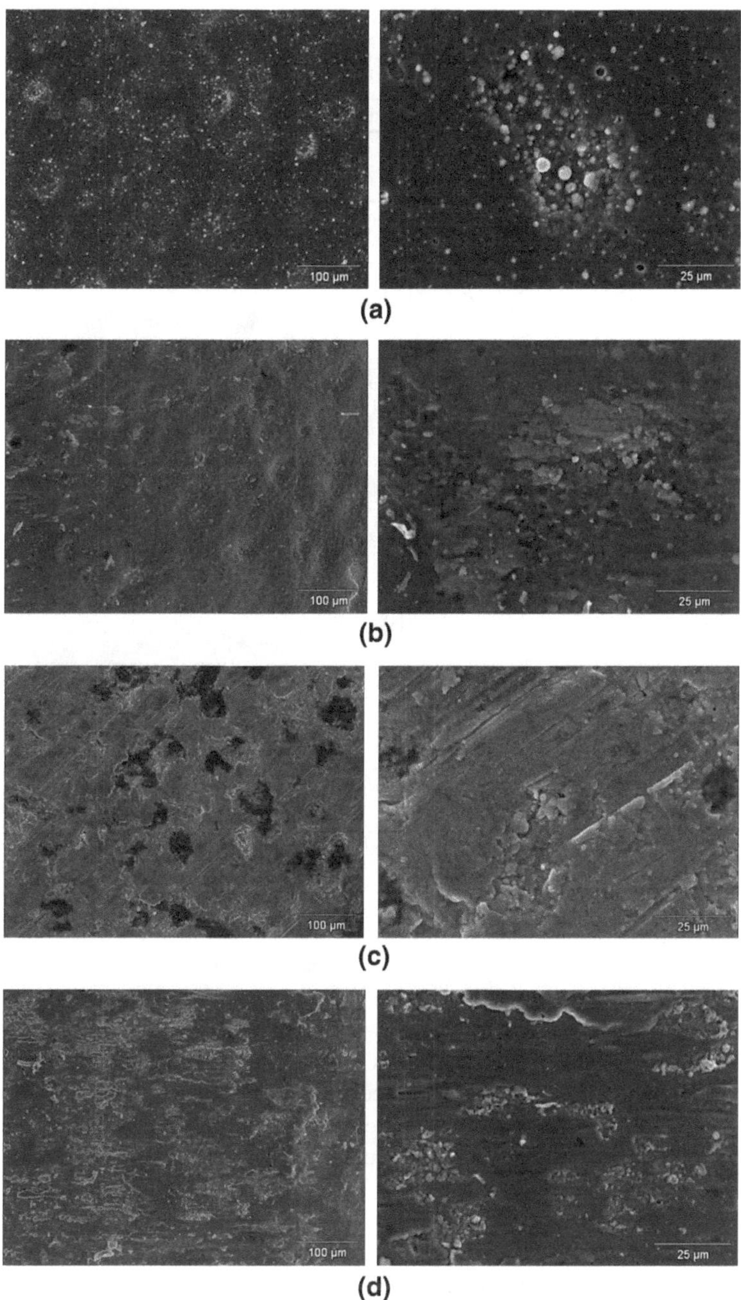

Fig. 3.11 Hot forging dies surfaces before and after one hundred strokes: **a**, **b** PN-PVD; **c**, **d** PN-PO (Scanning Electron Microscopy)

Table 3.1 Vickers hardness numbers (5 N, 20 s) of forging dies coated by PN-PVD and PN-PO before and after one hundred forging strokes

Coating	Before forging		After forging	
	Mean	Standard deviation	Mean	Standard deviation
PN-PVD	1277	190	1024	156
PN-PO	897	205	797	25

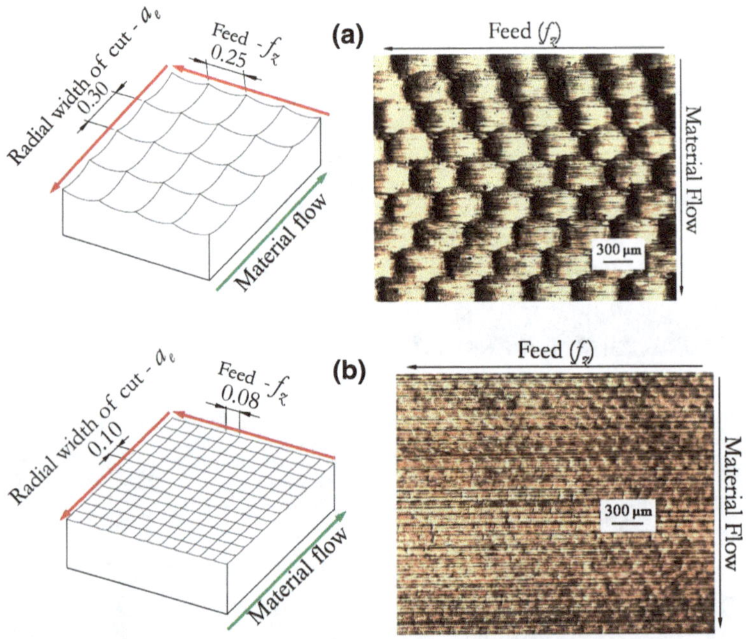

Fig. 3.12 Roughness profiles of two machining conditions before forging tests

is stored and dragged to establish the MPHDL and MPHSL mechanisms [44, 45], or to increase the friction coefficient between roller and steel sheet in cold rolling, as well to decrease sheet surface defects [46].

To analyze the influence of the surface texture on the wear mechanisms of hot forging dies, four conditions were applied to mill the flash land: radial width of cut—a_e (0.08, 0.10 and 0.30 mm) and feed-f_z (0.08 and 0.25 mm) [47]. Fig. 3.12 shows the conditions which present the best (condition #1—Fig. 3.12a) and the worst results (condition #2—Fig. 3.12b).

After 125 forging tests the dies were evaluated by light optical microscopy as shown in Fig. 3.13. The surface machined with condition #1 (Fig. 3.13a) presented the best results: the roughness profile was preserved and practically no abrasion or adhesion were observed. Otherwise condition #2 (Fig. 3.13b) presented the worst results, as the initial texture was completely modified by intense wear caused by adhesion and plastic deformation.

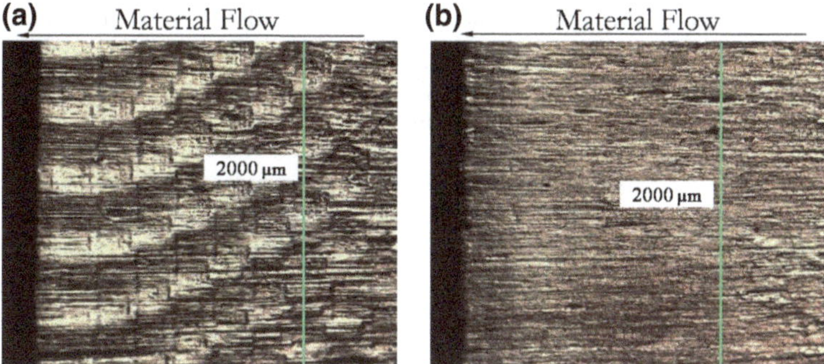

Fig. 3.13 Roughness profiles of two machining conditions after forging tests (Light optical microscopy)

The best tribological performance was obtained with condition #1, surface with microcavities generated by similar and high values of f_z and a_e. It enabled a larger amount of lubricant to be stored and dragged, reducing the friction between the surfaces.

The topography composed of microcavities with low values of f_z and a_e (condition #2) presented the worst tribological performance among all the tested conditions, probably because of its low ability to retain lubricant due to its small microcavities.

In condition #2, abrasion was very aggressive in the first fifty forging tests, but in the complete experiment with 125 tests, plastic deformation predominated due to the decreased hardness of the die's subsurface layers. Moreover, adhesion was higher than in condition #1.

Condition #2 also required a longer manufacturing time due to the low f_z an a_e values employed. On the grooved surfaces (f_z lower than a_e), the wear started on the grooves, reducing the roughness and impairing the retention of lubricant. This facilitated abrasion and adhesion, which were higher than in the condition with microcavities and high f_z and a_e values.

3.5 Conclusion

The global drives towards cost reduction, energy conservation, and to the use of non-polluting lubricants in metal forming processes challenges academic and industrial researchers to develop new and more effective lubricants, lubrication methods, and coating techniques. Therefore a lot of significant work has being done and a small part of it was referred in this chapter.

The comprehensive understanding of the basic concepts of tribological systems is essential to achieve competitive costs, high productivity, and environment-friendly lubricants. Future researches shall focus on advanced technologies like nanolubricants made with nanoparticles and nanotubes designed to fill the asperities of both surfaces in contact with very small and enough quantities of lubricant. Moreover, the time consuming and expensive surface coatings today in use will be substituted for nanostructured surfaces fabricated by laser and chemical processes, more effective and without disposal of hazard compounds.

References

1. Tselikov A (2003) Stress and strain in metal rolling. University Press of the Pacific, Berkeley
2. Ngaile G, Saiki H, Ruan L, Marumo Y (2007) A tribo-testing method for high performance cold forging lubricants. Wear 262(5–6):684–692
3. Grün F, Sailer W, Gódor I (2012) Visualization of the processes taking place in the contact zone with in situ tribometry. Tribol Int 48:44–53
4. Yanagida A, Kurihara T, Azushima A (2010) Development of tribo-simulator for hot stamping. J Mater Process Technol 210(3):456–460
5. Wang L, Zhou J, Duszczyk J, Katgerman L (2012) Identification of a friction model for the bearing channel of hot aluminium extrusion dies by using ball-on-disc tests. Tribol Int 50:66–75
6. Kim YS, Jain MK, Metzger DR (2012) Determination of pressure-dependent friction coefficient from draw-bend test and its application to cup drawing. Int J Mach Tools Manuf 56:69–78
7. Bowden FP, Tabor D (1986) The friction and lubrication of solids vol 1, 2nd edn. Frank Philip David Oxford University Press, Oxford
8. Panagopoulos CN, Georgiou EP (2010) Cold rolling and lubricated wear of 5083 aluminium alloy. Mater Des 31(3):1050–1055
9. Grob K, Marmiroli G (2009) Assurance of compliance within the production chain of food contact materials by good manufacturing practice and documentation—Part 3: lids for glass jars as an example. Food Control 20(5):491–500
10. Bosch MJ, Schreurs PJG, Geers MGD (2009) On the prediction of delamination during deep-drawing of polymer coated metal sheet. J Mater Process Technol 209(1):297–302
11. Caudle WM, Guillot TS, Lazo CR, Miller GW (2012) Industrial toxicants and Parkinson's disease. NeuroToxicology 33(2):178–188
12. Rao KP, Prasad YVRK, Xie CL (2011) Further evaluation of boric acid vis-à-vis other lubricants for cold forming applications. Tribol Int 44(10):1118–1126
13. Bay N, Azushima A, Groche P, Ishibashi I, Merklein M, Morishita M, Nakamura T, Schmid S, Yoshida M (2010) Environmentally benign tribo-systems for metal forming. CIRP Ann—Manuf Technol 59(2):760–780
14. Caminaga C, Neves FO, Gentile FC, Button ST (2007) Study of alternative lubricants to the cold extrusion of steel shafts. J Mater Process Technol 182(1–3):432–439
15. Gariety M, Ngaile G, Altan T (2007) Evaluation of new cold forging lubricants without zinc phosphate precoat. Int J Mach Tools Manuf 47(3–4):673–681
16. Arentoft M, Bay N, Tang PT, Jensen JD (2009) A new lubricant carrier for metal forming. CIRP Ann—Manuf Technol 58(1):243–246
17. Wilson WRD, Walowit JA (1971) An isothermal hydrodynamic lubrication theory for hydrostatic extrusion and drawing processes with conical dies. J Lubr Technol 93(1):69–74
18. Kucharski S, Starzyński G, Bartoszewicz A (2010) Prediction of surface roughness in metal forming with liquid lubricant. Tribol Int 43(1–2):29–39
19. Sun J, Lu W, Ma Y, Shi Q, Zhang A, Li J (2008) Surface quality of cold rolling aluminum strips under lubrication condition. J Univ Sci Technol Beijing 15(3):335–338
20. Button ST (2011) Numerical and experimental analysis of lubrication in strip cold rolling. J Brazilian Soc Mech Sci Eng 33(2):189–196

21. Martinez, GAS (1998) Lubrication in the wire drawing tribosystem at high drawing speeds, Doctorate Thesis, University of Campinas, Brazil, p 121
22. Le HR, Sutclife MPF (2000) Analysis of surface roughness of cold-rolled aluminium foil. Wear 244(1–2):71–78
23. Costa HL, Hutchings IM (2007) Hydrodynamic lubrication of textured steel surfaces under reciprocating sliding conditions. Tribol Int 40(8):1227–1238
24. Dick K, Lenard JG (2005) The effect of roll roughness and lubricant viscosity on the loads on the mill during cold rolling of steel strips. J Mater Process Technol 168(1):16–24
25. Lenard JG (2002) Metal forming science and practice: a state-of-the-art in honour of professor J.A. Schey's 80th birthday. Elsevier Science, Amsterdam
26. Wiklund D, Rosén BG, Wihlborg A (2009) A friction model evaluated with results from a bending-under-tension test. Tribol Int 42(10):1448–1452
27. Costa HL, Hutchings IM (2009) Effects of die surface patterning on lubrication in strip drawing. J Mater Process Technol 209(3):1175–1180
28. Azushima A (2006) In situ 3D measurement of lubrication behavior at interface between tool and workpiece by direct fluorescence observation technique. Wear 260(3):243–248
29. Lavtar L, Muhič T, Kugler G, Terčelj M (2011) Analysis of the main types of damage on a pair of industrial dies for hot forging car steering mechanisms. Eng Fail Anal 18(4):1143–1152
30. Ebara R, Kubota K (2008) Failure analysis of hot forging dies for automotive components. Eng Fail Anal 15(7):881–893
31. Santaella ML (2009) Influence factors on forging die wear of automotive connecting rods, Master Dissertation, University of Campinas, Brazil, p 120
32. Terčelj M, Smolej A, Fajfar P, Turk R (2007) Laboratory assessment of wear on nitrided surfaces of dies for hot extrusion of aluminium. Tribol Int 40(2):374–384
33. Vilaseca M, Molas S, Casellas D (2011) High temperature tribological behaviour of tool steels during sliding against aluminium. Wear 272(1):105–109
34. Khader I, Hashibon A, Albina JM, Kailer A (2011) A wear and corrosion of silicon nitride rolling tools in copper rolling. Wear 271(9–10):2531–2541
35. Galakhar AS, Gates JD, Daniel WJT, Meehan PA (2011) Adhesive tool wear in the cold roll forming process. Wear 271(11–12):2728–2745
36. Paschke H, Weber HM, Kaestner P, Braeuer G (2010) Influence of different plasma nitriding treatments on the wear and crack behavior of forging tools evaluated by Rockwell indentation and scratch tests. Surf Coat Technol 205(5):1465–1469
37. Tillmann W, Vogli E, Momeni S (2010) Improvement of press dies used for the production of diamond composites by means of DUPLEX-PVD-coatings. Surf Coat Technol 205(5):1571–1577
38. Leskovšek V, Podgornik B, Jenko M (2009) A PACVD duplex coating for hot-forging applications. Wear 266(3–4):453–460
39. Podgornik B, Zajec B, Bay N, Vižintin J (2011) Application of hard coatings for blanking and piercing tools. Wear 270(11–12):850–856
40. Navinšek B, Panjan P, Gorenjak F (2001) Improvement of hot forging manufacturing with PVD and DUPLEX coatings. Surf Coat Technol 137(2–3):255–264
41. Gu ST, Chai GZ, Wu HP, Bao YM (2012) Characterization of local mechanical properties of laser-cladding H13–TiC composite coatings using nanoindentation and finite element analysis. Mater Des 39:72–80
42. Garrido AH, González R, Cadenas M, Battez AH (2011) Tribological behavior of laser-textured NiCrBSi coatings. Wear 271(5–6):925–933
43. Zhou S, Zeng X, Hu Q, Huang Y (2008) Analysis of crack behavior for Ni-based WC composite coatings by laser cladding and crack-free realization. Appl Surf Sci 255(5):1646–1653
44. Wagner K, Völkl R, Engel U (2008) Tool life enhancement in cold forging by locally optimized surfaces. J Mater Process Technol 201(1–3):2–8
45. Schubert A, Neugebauer R, Sylla D, Avila M, Hackert M (2011) Manufacturing of surface microstructures for improved tribological efficiency of power train components and forming tools. CIRP J Manuf Sci Technol 4(2):200–207

46. Zhitong W, Mingjiang Y (2011) Laser-guided discharge texturing for cold mill roller. J Mater Process Technol 211(11):1678–1683
47. Magri ML (2011) Influence of the surface topography on the life of hot forging dies, Master Dissertation, University of Campinas, Brazil, p 135

Chapter 4
Tribology in Hot Rolling of Steel Strip

D. B. Wei and Z. Y. Jiang

Abstract Contact friction is of crucial importance for accurate simulation, optimum design and control of industrial rolling processes. It affects the shape, profile, dimensional accuracy and surface quality of hot rolled strips. This chapter focuses on the tribology of hot strip rolling of plain carbon steel and stainless steel, which is significantly affected by oxide scales. The fundamental of oxidation of pure iron, plain carbon steel and stainless steel, and the formation of oxide scales in hot rolling process have been discussed. The morphology of the oxide scales and their deformation behaviours that depend on oxide scale thickness, constitution and the rolling parameters have been disclosed. Surface roughness of oxide scales and the tribological effect of oxide scales in hot strip rolling have been studied. A multi oxide scale layers simulation has been established to study the deformation and fracture of oxide scales taking into account the effect of surface roughness.

4.1 Introduction

In rolling process, the contact friction is of crucial importance for the accurate modeling, optimum design and control of industrial rolling processes. As it has a relationship with the rolling force and rolling pressure distribution, friction affects

D. B. Wei · Z. Y. Jiang (✉)
School of Mechanical, Materials and Mechatronic Engineering,
University of Wollongong, Wollongong, NSW 2522, Australia
e-mail: jiang@uow.edu.au

D. B. Wei
e-mail: dwei@uow.edu.au

J. P. Davim (ed.), *Tribology in Manufacturing Technology*, Materials Forming, Machining and Tribology, DOI: 10.1007/978-3-642-31683-8_4,

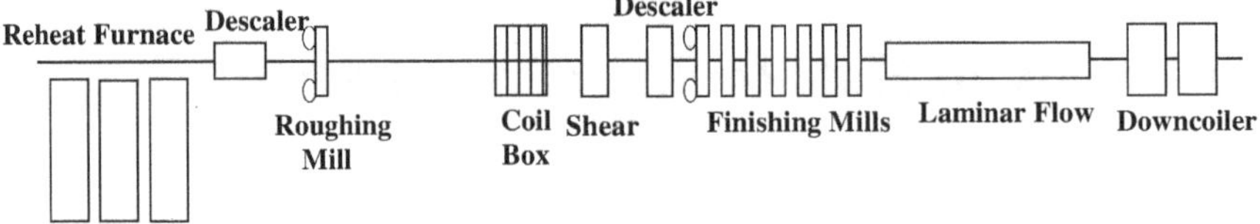

Fig. 4.1 Layout of a production line of hot rolling of steel strip

the roll gap and roll wear. As a result, it affects the shape, profile, dimensional accuracy and surface quality of the steel strip.

A well-defined boundary condition is vital for generating an accurate model for simulating any metal manufacturing process [1–5]. It is important to characterize the features of the oxide scale in hot strip rolling of steels because the oxide scale on the strip surface affects friction and thermal conductivity coefficient. The oxide scale layer influences the heat transfer at the interface between the strip and the roll [1, 6]. The thermal conductivity of oxide scale is a factor of about 10–15 times less than that of steel [7]. It also influences the rate of heat loss between rolling passes. Consequently, the rolling force and the friction condition depend on the thickness and the microstructure of the oxide scale.

A typical production line of hot rolling of steel may comprise walking beam reheat furnace, a reversible roughing rolling mill with vertical rolls, a hot coil box, a flying shear, a pair of small vertical guide roll, a 7-stand finishing rolling mill and two downcoilers, as shown in Fig. 4.1.

Surface roughness is a core quality feature of hot rolled steel strip. The order of magnitude of the influence of oxide scale layers on surface roughness has been investigated.

During the hot strip rolling process of steels, thick oxide scales (primary scale) formed on the slabs during reheating are removed by a hydraulic descaler prior to the roughing process. After 3–7 passes of rolling in the roughing mill, more oxide scales (secondary oxide scale) formed on the steel surface during rough rolling are removed by a high-pressure steam descaler immediately after the roughing rolling mill, and the other high-pressure steam descaler immediately before the finishing mill. Oxide scale formed during finishing rolling (tertiary oxide scale) is retained to room temperature after laminar flow cooling and coiling [8].

The deformation behavior of oxide scale is an important subject in hot strip rolling, which may depend on oxide scale thickness, constitution and rolling parameters [4, 9–11]. The mechanical properties of oxide scale are attributed to temperature [11]. As the oxide scale formed during finishing process is not fractured but adheres well to the steel substrate, it is believed to be deformable during finishing rolling. The thickness of oxide scale is repeatedly reduced at each rolling stand in the same proportion as that of the steel substrate. The oxide scale thickness grows while the steel strip is travelling between adjacent stands. Therefore, the oxide scale thickness is determined by three factors: strip surface temperatures, strip speeds through the finishing mills and rolling reductions [8].

The deformation or fracture of oxide scale affects the interface between the roll and the strip [12].

Although several descaling approaches have been proposed, they cannot always remove the oxide scale before rolling [12]. In practice, oxide scale is playing a growing importance in determining the friction in hot strip rolling and the surface quality of hot strip.

4.2 Structure of Oxide Scale and Oxidation Kinetics in Pure Iron and Carbon Steel

Oxide scales generally grow by a mechanism whereby positive metal ions form at the base metal surface and diffuse through the already formed oxide lattice until they reach the oxide scale/atmosphere interface, where reaction with oxygen occurs. Oxide scale growth thus occurs in an "inside out" fashion, with inner layers of oxide scale being formed prior to the outer layers [13].

4.2.1 Pure Iron

The reactions between iron and oxygen are exothermic so an over-temperature phenomenon is present during the initial oxidation stage. Despite the rapid initial reactions, the longer-term oxidation rate of iron under isothermal-oxidation conditions is quite steady and usually follows the parabolic law [13].

The oxidation of pure iron displays different behaviour in three different temperature ranges, which are 700–1,250 °C, 570–700 °C and below 570 °C.

The oxide scale developed at 700–1,250 °C follows the parabolic law and comprises an extremely thin outermost hematite (Fe_2O_3) layer, a thin intermediate magnetite (Fe_3O_4) layer, and a thick inner wustite (FeO) layer. The diffusion coefficient of iron in wustite is much greater than in magnetite and that the diffusion of oxygen and iron through the hematite layer is extremely slow [14].

The isothermal-oxidation kinetics of pure iron in air or oxygen during steady-state reaction conditions at this temperature range can be expressed by the Tammann-type parabolic equation [13]

$$dx/dt = k_x'/x$$

or

$$x^2 = k_x t + x_0^2 \tag{4.1}$$

where x denotes the total scale thickness, t the oxidation time, x_0 the initial oxide scale thickness at the start of the parabolic oxidation stage, and k_x the parabolic rate constant, usually measured in cm^2s^{-1}. In the original Tammman equation, x_O was taken as 0. The parabolic rate constant is exponentially dependent on temperature, expressed as [13]

$$k_x(\text{cm}^2\,\text{sec}^{-1}) = k_x^0 \exp(-Q/RT) \tag{4.2}$$

where $k^0{}_x$ is a constant, Q the activation energy of iron oxidation in J-mol^{-1}, T absolute temperature, and R the gas constant in J-K^{-1} mol^{-1}. The activation energy Q of iron oxidation is typically found to be equivalent to the activation energy of iron diffusion in wustite.

Paidassi [15] found that the thickness ratios of hematite, magnetite, and wustite layers in the oxide scale samples formed isothermally at 700–1,250 °C were nearly constant at about 1:4:95.

The oxidation behavior of iron at 570–700 °C is more complex and the reported results are less consistent [13]. Different thickness ratios of the oxide scale layers were observed. However, it was agreed that the isothermal-oxidation kinetics followed the parabolic law.

Below 570 °C, the wustite is thermodynamically unstable. The scale comprises an outer hematite layer and an inner magnetite layer [16].

4.2.2 Carbon Steel

Because various alloying and impurity elements present in steels, the oxidation behaviour and the oxide scale structure developed are more complex than that for pure iron.

Although the reported kinetics data for steel oxidation in air and oxygen are sparse, they generally follow the parabolic law. Most of them were derived from oxidation of low-carbon steels. The oxidation kinetics of mild steel was compared with those of pure iron in air and oxygen at 700–1,100 °C [13]. It was found that the oxidation kinetics followed the parabolic law and the oxidation rates of mild steel were lower than those of pure iron in either oxygen or air. The oxidation kinetics for steels having similar compositions at 730–935 °C for 60 min and at 1,050 °C for 30 min in air follows the parabolic law [17]. The oxygen level significantly affects the initial oxidation rates, but has no effect on the subsequent parabolic oxidation rates [18]. The oxidation of carbon steel displays different behaviour above and below 800 °C.

4.2.2.1 800 °C and Above

The scale structures formed on carbon steel in air or oxygen after oxidation under short time exposure are usually similar to those formed on pure iron, comprising a thick inner wustite layer, a thin magnetite layer and a very thin surface hematite layer [19]. After slightly longer time exposure in air, blisters start to form, and the oxide scale structure becomes irregular [19].

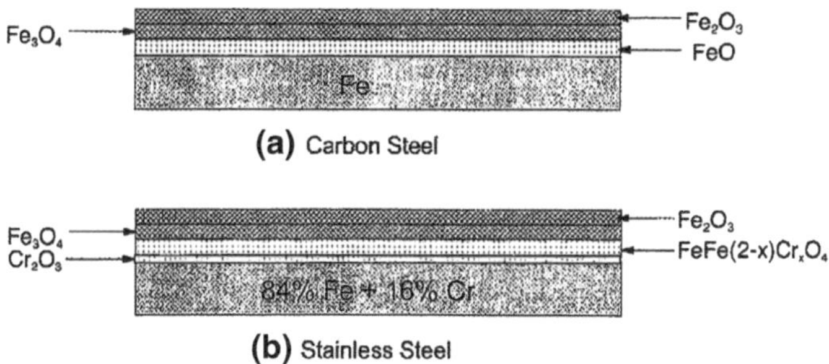

Fig. 4.2 Structural diagrams of oxide scale [21]

Blisters form if the following conditions are met. First, stress is generated in the oxide scale. Second, the oxide scale has sufficient plasticity. Third, the oxide scale-steel interface adhesion is weakened at certain locations, which could be the result of vacancy condensation [20]. Finally, a gas phase accumulated inside the blisters to prevent them from collapsing or healing after they have formed [13]. Because iron supply at blistered area is terminated as a result of oxide scale detachment from the substrate, and the wustite formed earlier is oxidized to higher oxides at the later stages, the blistered areas reveal a thin oxide scale layer comprising primarily of hematite and magnetite arching outward over a cavity between the oxide scale and the substrate [19].

The average oxide scale thickness after oxidation in air for a certain period is found to reach a maximum at a temperature around 980–1,010 °C. Above 980 °C, oxide scale blistering may occur, which results in the average oxide scale thickness decreasing with increasing oxidation temperature [13]; the thickness becomes the lowest at about 1,100 °C but increases again at temperatures above 1,100 °C.

4.2.2.2 Below 800 °C

After exposure for 1 h or longer (up to 16 h) at 580–760 °C, the oxide scales formed on AISI 1006 steel in its hot-rolled and pickled conditions was irregular in their thickness and structures, comprising primarily of hematite and magnetite, with very little or no wustite [19]. The oxide scale structures formed at 450–560 °C is similar to those formed on pure iron with a two-layer hematite-magnetite structure [19].

4.2.3 Structure of Oxide Scale and Oxidation Kinetics in Stainless Steels

The relative affinities of metals to oxygen are: silicon > manganese > chromium > iron > nickel [21]. Therefore, in an iron-based alloy containing chromium as a solute, an outer layer of chromium oxide Cr_2O_3 will preferentially form during the early stages of oxidation prior to the formation of any iron oxides. Although iron thermodynamically has much less affinity for oxygen than chromium does, it still oxidized at a much faster rate than chromium, because chromium oxide layers provide a far higher degree of protection against further oxygen pickup than iron oxide layers [21].

Conventional stainless steels are grouped within the chromia-forming alloys, which form a Cr_2O_3 layer that protects the bulk against corrosion up to 627 °C. Outward diffusion of cationic species and inward diffusion of oxide ions are responsible for the formation of the protective layer. As a consequence of these diffusion mechanisms, a parabolic rate law fits the oxidation process [22].

The two major potential components of a stainless steel oxide scale are M_2O_3 rhombohedral phase and M_3O_4 spinel phase [21]. Because iron and chromium are the predominant metals in the oxide scale layers of conventional stainless steels, previous discussion focused on the oxides of the two metals. The classical chemical compositions and structures of oxide scale in carbon steel and stainless steel are compared and shown in Fig. 4.2.

Generally, the oxide growth rate is parabolic, which is indicative of an oxide scale layer that protects the underlying alloy from further oxidative attack. This oxide scale layer can abruptly rupture due to mechanical stresses and lead to a period of more rapid oxidation. This is generally referred to as "breakthrough" oxidation. The more rapid oxidation after breakthrough is likely due in part to Cr-depletion of the underlying alloy surface, since Cr-rich oxide scales are formed during the early stages of the heating cycles [21].

304 stainless steel develops a Cr_2O_3 protective layer at temperature $< =$ 900 °C. As a result of solid state diffusion mechanisms, a parabolic rate law is generally followed [23]. After oxidation at 1,000 °C in air for 10 h [24], the composition of the oxide layer of in 304 stainless steel reveals either a multiplayer structure with an internal oxide scale composed of Cr_2O_3, $Mn_{1.5}Cr_{1.5}O_4$ and an external scale constituted of iron oxides. The growth of $FeCr_2O_4$ and $(Cr, Fe)_2O_3$ oxides after 9 h at 1,000 °C was also studied [23].

A study [25] on the oxidation of 304L steel oxidized at 1,000 °C in air was carried out. A transient state up to 9 h results in a layer not completely established made up of Cr_2O_3 and phases of the spinel oxides $NiCr_2O_4$ and $MnNi_2O_4$. After 9 h, the oxidation follows a parabolic rate law corresponding to the formation of hematite and mixed oxide $(Fe, Cr)_2O_3$.

Two main possibilities to increase the resistance to oxidation at high temperature of stainless steels are to increase Ni, Cr and to increase the reactive element (RE) effect by rare earth elements. RE effects include: an improvement in oxide scale adhesion or resistance to spallation, a change in the oxide scale-growth mechanism, a reduction in the oxidation rate and a modification of the oxide scale microstructure.

4.2.4 Oxidation of Stainless Steels in Moist Atmosphere

Many studies focus on the oxidation behavior in hot strip rolling for plain carbon steel, but few studies have been carried out directly on the oxidation behaviour in hot strip rolling of stainless steels. Although the oxidation time in most study on the oxide behavior of stainless steels is much longer than the time required for the hot strip rolling, some results are still helpful. Those studies on the oxidation behaviour of stainless steels under short time exposure, in moist atmosphere or during hot rolling are highlighted.

The presence of water vapour changes the oxidation behavior of metals and alloys compared to the dry oxygen case by increasing the corrosivity of the environment [26]. Generally, the exposure of Fe–Cr alloys to O_2/H_2O mixtures in the temperature range of 500–900 °C can result in Cr depletion of the protective oxide by evaporation of chromium species $CrO_2(OH)_2$ [27–31]. This leads to nucleation of non-protective Fe-rich oxide on the surface, which results in an increased oxidation rate.

A general rule is the oxidation rate of a metal increases with increasing temperature. In moist atmospheres, a relationship between the oxidation rate and the temperature is sometimes not obeyed, which results in an abnormal rapid oxidation. It was found [32] that an abnormal rapid oxidation occurred specifically at about 600 °C under a constant water vapor pressure $P_{H_2O} = 3.4 \times 10^3$ for $P_{O_2} = 2.0 \times 10^3$ Pa and below for 430 stainless steel. Water vapor affects not only the initiation of abnormal oxidation but also the growth of whiskers and nodules in the oxide scale.

A study [33] on the oxidation of 304 stainless steel in steam temperatures between 900 and 1,350 °C found that the oxidation kinetics was to obey the parabolic law during the first period of 8 min. After the first period, the parabolic reaction rate constant decreased in the case of heating temperatures between 1,100 and 1,250 °C.

Effect of water vapor on annealing scale formation of 316 stainless steel at 800–1,030 °C up to 8 h was studied [34]. A dramatic increase in weight gain was found if 0.1 atm water vapor was present in air. After being oxidized in wet air at 1,030 °C for 0.5 h, the scratch-type surface appearance implied the very thin oxide (2–3 μm), but exhibited good adherent characteristics with the matrix; for 1 h, nodule-like oxide observed; for 1.5 h, the nodule-like oxide disappeared but was replaced with the formation of a continuous surface oxide scale. A multi-layered oxide scale structure was found with prolonged exposures of 3, 5 or 8 h.

Evaporation of Cr-species for 304L at 600 °C in O_2/10 %H_2O was observed [28, 29]. The chromium concentration below the interface metal/oxide drops and reaches a critical value for breakaway. An evaporation-depletion mechanism was presented [35].

The oxidation behavior of 304 and 439 stainless steels at 850–950 °C under various oxygen pressures, and in the presence or not of water vapour for 50 h was compared [13]. For 439 stainless steel, the kinetics follows a parabolic law in all cases, indicating that the film growth is controlled by diffusion. The oxide scale is

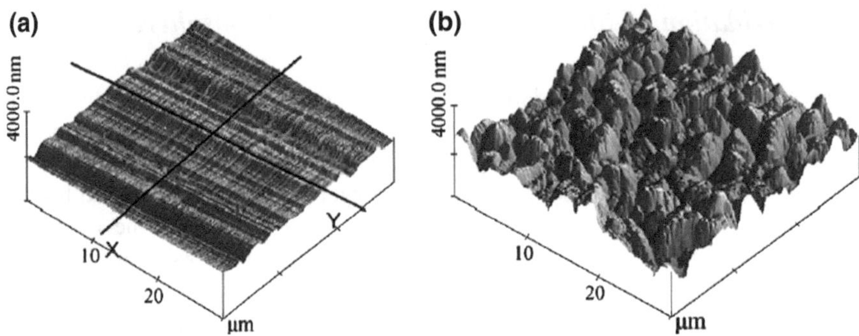

Fig. 4.3 Image of oxide scale surface of low carbon steel [39]. **a** Original surface **b** holding 80 s in air at 800 °C

mainly made up of Cr_2O_3. At the lowest temperature in the presence of water vapour, whiskers are observed. For 304 stainless steel, a parabolic kinetics was observed only at 850 °C, while a deviation to the parabolic behavior is observed above 850 °C [36].

4.2.5 Oxidation of Stainless Steels in Hot Strip Rolling

It was found [37] that the oxide scale formed on 304 hot rolled strip contains spinel constituents like $FeCr_2O4$, $NiFe_2O_4$, Fe_3O_4 and Fe_2O_3 while the oxide scale formed on 430 hot rolled strip consists of $FeCr_2O_4$, Fe_3O_4 and Fe_2O_3.

4.3 Characteristics of Oxide Scale Layer

The characteristics of oxide scale are very significant for the surface quality of steels. It is also important for creating an accurate numerical model to simulate the behaviour of oxide scale deformation.

4.3.1 Surface Characteristics

The very detailed surface morphology of oxide scale during hot strip rolling was discovered using Atomic Force Microscopy (AFM) [38–40]. One of the 3D image of the surface of low carbon steel sample before and after 80 s oxidation at 800 °C is shown in Fig. 4.3. It can be seen that the surface has obvious machined texture before oxidation. The surface roughness exhibits primary, secondary, and tertiary wavelengths [38]. After oxidation, the machined texture disappeared and a layer of oxide scale covers the surface, and there is no clear secondary or tertiary

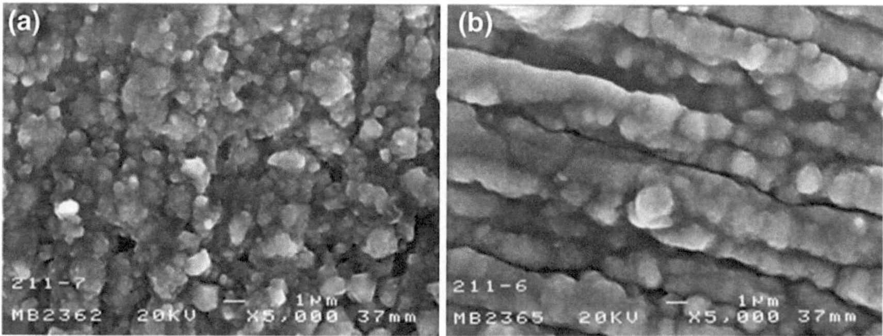

Fig. 4.4 Surface morphology of a low-nickel austenitic stainless steel after exposure for 500 h at (*left*) 700 °C and (*right*) 600 °C [41]

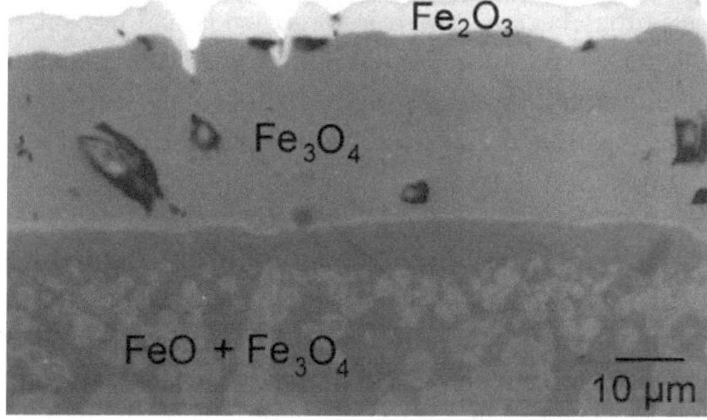

Fig. 4.5 Three layered oxide scale of carbon steel exposure at 1,000 °C for 0.5 h in humid air (19.5 vol. %) [42]

wavelength. The asperity profile becomes similar while the surface becomes coarser. A typical oxide scale wavelength is about 20–30 μm. The average oxide scale surface roughness is in a range of 2–4 μm after descaling.

A study [41] shows the oxide scale surface morphology on low-nickel austenitic stainless steel (main chemical composition: Cr 17.02, Mn 10.58, Ni 2.63, Cu 1.97, Fe balance, wt %) by SEM. At high temperature, the surface morphology may be as similar as that on carbon steel, while at low temperature, the original grinding texture is well conserved even after a long time due to the smaller oxide scale thickness, as shown in Fig. 4.4. The surface-oxide composition rich in Mn and Cr, and the phase is Mn_2O_3 and $MnCr_2O_4$.

4.3.2 Morphology on Cross Section

The structure of the oxide scale on the surface of carbon steel is complex and it may include two or three layers, as shown in Fig. 4.5. The oxide scale layer may

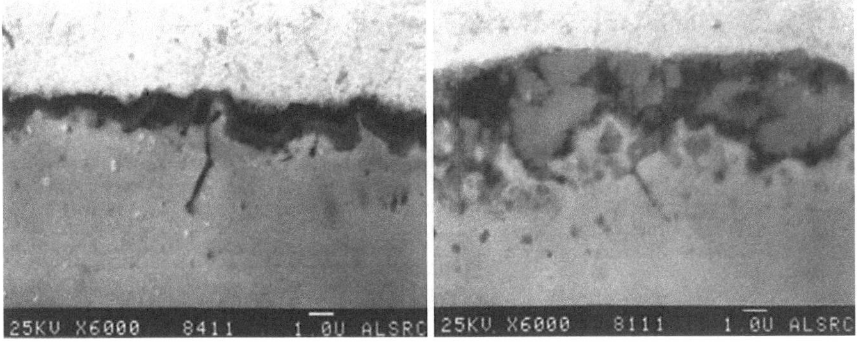

Fig. 4.6 Cross-section of scale on 304 stainless steel strip exposure at 1,176 °C after 10 s (*left*) 7 % O_2; (*right*) < 0.2 % O_2 [45]

be porous because there is large voids and crystal size on the order of the scale thickness. The porosity of the oxide scale depends on the cooling rate. High cooling rate may lead to larger pores and cracks in the entire oxide, while cracks occur near the metal/oxide interface in the case of low cooling rate [42]. However, if thickness of oxide scale is small, the oxide scale layer usually is relatively compact [19, 43, 44].

In the case of oxidation of stainless steels under short time exposure, the thickness of oxide scale is small [45]. Figure 4.6 shows the SEM cross-section view of oxide scale on 304 stainless steel strip at 1,176 °C after 10 s TAT (time at temperature) anneal in < 0.2 and 7 % O_2. At the very low oxygen levels, the oxide scale formed is discontinuous and islands of oxides while a more or less continuous and uniform oxide scale was formed at the higher oxygen level of 7 %. X-ray Diffraction gave the following constituents in oxide scale: Mn_3O_4, Fe_3O_4, $FeCr_2O_4$, $Mn_{1.5}Cr_{1.5}O_4$ and $NiMn_2O_4$.

4.4 Mechanical Properties and Thermal Properties of the Oxide Scale

4.4.1 Mechanical Properties and Measurement Methodology

The mechanical properties of oxide scales are important for the surface quality of hot rolled steel strip. The mechanical data can help optimisation of hot rolling process [42]. It is also the base of numerical simulation of hot strip rolling, in which the deformation behaviour of oxide scale is considered.

The most important attributes of the oxide scale in metal forming are its hardness and yield strength, which indicate whether the oxides are abrasive. The hardness of hematite, magnetite and wustite are 460, 540 and 1,050 Hv respectively at room

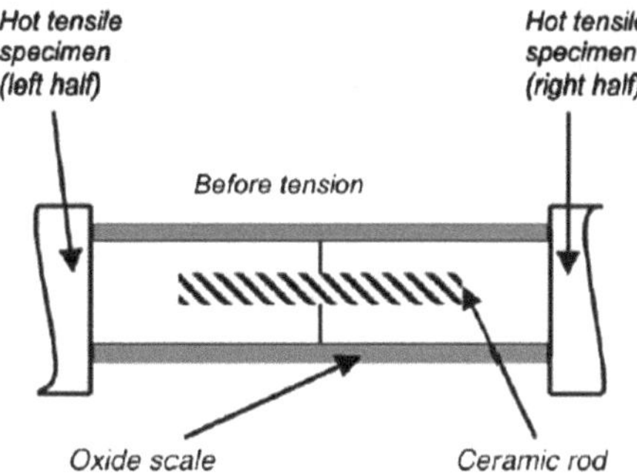

Fig. 4.7 Tensile specimen for evaluating scale adhesion and fracture loads [47]

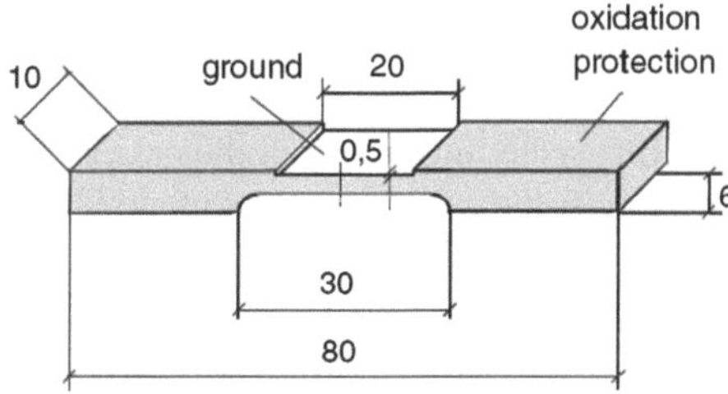

Fig. 4.8 Four-point-bend test specimen [42]

temperature [2]. The hardness of the oxides is temperature dependent [11]. The above values are 105, 366 and 516 HV respectively at 900 °C [46].

The measurement of the mechanical properties of oxide scale, which requires a combination of experiments under appropriate operating conditions and computer-based modelling [42] is a challenge. Especially, the measurement at high temperature is quite difficult. Besides some useful results, the methodologies are briefly reviewed here.

A hot tensile test was designed by Krzyzanowski et al. [47]. Round tensile specimen was cut into two equal parts. The two parts were connected together by a ceramic pin and the oxidized surfaces match each other accurately, as shown in Fig. 4.7. The specimens were oxidized in a vertical cylindrical induction furnace. A small compressive load was applied to allow a continuous oxide scale layer formed and prevent the oxidation on the joint face. After the slow cooling rate, tensile test was operated to obtain those parameters. The separation load measured without oxidation was registered for all tests as a background of bonding. Subtracting the separation load measured without oxidation from the separation load measured with oxidation to obtain the oxide scale effect.

Thermally-grown oxide scales 4-point bend test was conducted by Echsler et al. [42]. The specimen developed for applying a defined compressive-stress state in the metal/oxide system is shown in Fig. 4.8. After an aluminide-diffusion coating of all surfaces, the upper surface in the middle part was ground, so that the investigated iron oxide scales only form on the ground part. The four-point bend fixtures were placed in a vertical tube furnace installed on a modified four-point-bend test facility. The oxide scales formed on mild steel at 800, 900 and 1,000 °C in dry air, humid air and laboratory air. The test was suitable for oxide-scale

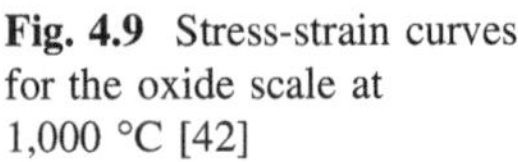
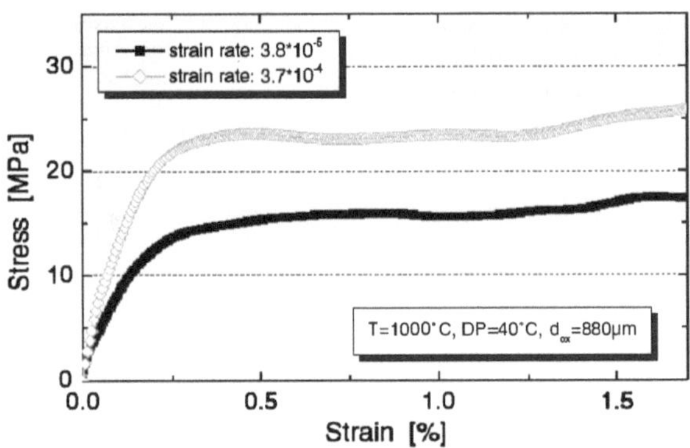

Fig. 4.9 Stress-strain curves for the oxide scale at 1,000 °C [42]

thickness values equal to or higher than 100 μm because detachment or spallation of oxide scale may occur in the case of lower oxide scale thickness. In the case of humid air, the water vapor contents 7.0, 12.0 and 19.5 % are relevant in the hot rolling process. It was found that the water vapor does not have any significant effect on the creep properties of the oxide scales. The stress–strain curve for oxide scales at 1,000 °C is shown in Fig. 4.9.

Hidaka [48] studied the deformation of FeO oxide scales upon tensile tests at 600–1,200 °C. In their tests, FeO was obtained by complete oxidation of pure iron. It was found that FeO oxide scale deformed plastically at above 700 °C, elongation over 100 % occurred and steady-state deformation was obvious at above 1,000 °C.

Riedel [49] proposed a model for the deformation and cracking of second-phase layers on creeping metals, which took into account elastic and creep deformation. Hancock [50] described a method to determine the dimensions of a "composite defect" on surface oxide scales by summing the actual defects present in the oxide scale using well-established techniques of linear elastic analysis. Schutze [51] developed an approach to model the oxide scale fracture by calculating the stress situation in the oxide/metal system.

4.4.2 Thermal Properties and the Effect of Oxide Scale on Heat Transfer

Both thermal contact and friction are boundary conditions to the hot rolling process and to determine the strip flow behavior in the roll gap [52]. The thermal conduction at roll/strip interface represents the majority of the heat loss of the strip. The situation is complicated by the presence of oxide scale, which changes the contact state, and has a critical influence on the thermal resistance and heat flux at the interface [53].

The magnitude of heat transfer through the contact interface between the roll and the 316 stainless steel was described by an effective interfacial heat transfer coefficient (IHTC) $C_e = q_e/\Delta T_{int}$, where q_e is the effective heat flux (kW/m^2), ΔT_{int}

Table 4.1 Thermophysical parameters of wustite, magnetite and hematite [1]

	K (W/m K)	C_p (J/kg K)	Density (kg/m^3)
Wustite	3.2	725	7,750
Magnetite	1.5	870	5,000
Hematite	1.2	980	4,900

the temperature difference between the roll and the strip, and the unit of C_e is kW/m^{2}K. For 316 stainless steel slab of initial thickness 25 mm, in the case of oxide scale thickness 0.08 and 0.15 mm, $C_e = 12.0$ and 8.0 kW/m^{2}K respectively if the reduction was 20 %, while $C_e = 25.0$ and 8.2 kW/m^{2}K respectively if the reduction was 40 %. Both oxide scale thickness and percentage reduction have considerable influence on the IHTC value. With an increase of oxide scale thickness, the IHTC values decrease obviously, whatever the percentage reduction is used [53]. With an increase of rolling reduction, the effective IHTC increases significantly due to the improved boundary contact and fresh steel flow into the oxide scale cracks [54]. The chemical composition of steel can affect the IHTC value by producing different quantity and density of the oxide scale on the surface of steel slab and altering rolling pressure [54].

Torresa et al. [1] took into account the different thermophysical properties of wustite, magnetite and hematite and developed a heat conduction model to predict the temperature distribution within the oxide layer on the surface of hot rolled carbon strip. The adopted thermal conductivity k and the heat capacity C_p of wustite, magnetite and hematite are shown in Table 4.1. However, the data of chromia is not available.

4.5 Evolution of Oxide Scale During Hot Rolling

4.5.1 Oxide Scale Growth and Deformation During Finishing Rolling

Oxide scale thickness is an important parameter for evaluating the behaviour of oxide scale in the roll bite because of its influence on the coefficient of friction and mechanical properties [9, 55]. The oxide scale thickness varied as the strip travels through the finishing mill. The oxide scale thickness formed on the transfer bar between the descaler and stand 1 depends on the strip temperature and rolling speed. As an example, in the case of five finishing rolling stands for hot carbon steel strip [8], it is over 20 μm before entering stand 1, then reduced between 10–15 μm in stand 1, about 5 μm of oxide scale forms on the strip during travelling between stands 1 and 2, which is determined by the strip temperature and rolling speed again. After repeated thickness reductions and growths through stands 2–5, the oxide scale thickness is reduced to less than 10 μm at the exit of the finishing mill, as shown in Fig. 4.10.

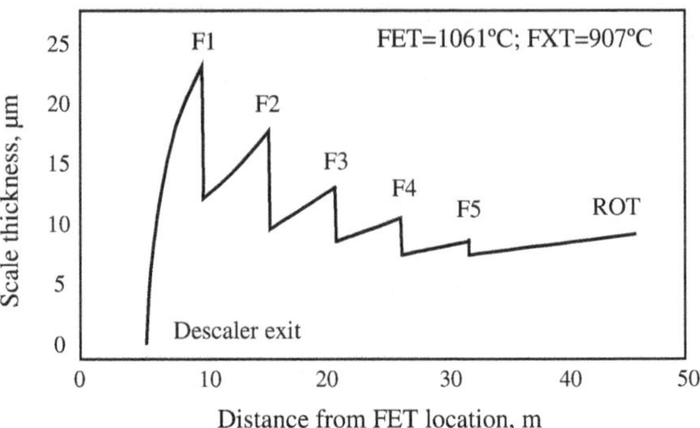

Fig. 4.10 Illustration of oxide scale growth in a five stands finishing rolling mills for carbon steel hot rolling strip [8], FET-finishing-mill entry temperature, FXT-finishing-mill exit temperature, ROT-run-out table

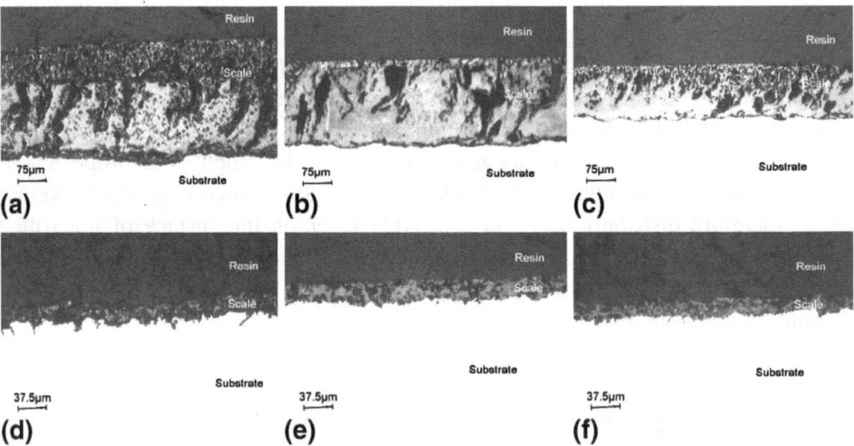

Fig. 4.11 Microstructures of the primary and secondary oxide scales of low carbon steel before and after deformation [56]. **a** Non-deformed primary scale; **b** primary scale deformed 15.8 % at rolling speed of 0.12 ms^{-1}; **c** primary scale deformed 33.7 % at rolling speed of 0.48 ms^{-1}; **d** non-deformed secondary scale; **e** and **f** secondary scale deformed 15.8 % at rolling speed of 0.24 and 0.72 ms^{-1} respectively

4.5.2 Oxide scale Structures Developed on Hot Rolled Steel Strip

The evolution and mechanism of the oxide scale structures developed on hot-rolled carbon steel strip were studied [44]. The oxide scale structures after finishing hot-rolling and cooling on the run-out table had a three-layer structure, which are the surface hematite layer, intermediate magnetite layer and inner wustite layer. The thickness of the magnetite layer was similar to that of the wustite layer, and the total oxide scale thickness varied with the finishing rolling temperature. There are no similar results for hot rolled stainless steel strip.

4.5.3 Deformation and Fracture Behavior of Oxide Scale Under Hot Rolling Conditions

Oxide scale is involved in high normal compressive stress from the work roll, tension stress due to elongation of the steel, and shear stress between the roll and the bulk material in the roll bite where the oxide scale is supposed to exhibit some deformation.

4.5.3.1 Carbon Steel

The effect of hot rolling conditions on the deformation behaviour of the oxide scale of low carbon steel has been studied [56–58]. The oxide scale thickness of both primary and secondary oxide scale decreases with bulk reduction in a linear relationship. The magnitude of deformation of primary oxide scale is around 8–10 times of that of secondary oxide scale, as shown in Fig. 4.11.

Temperature and lubrication condition have significant effect on the deformability of the oxide scale while the effect of rolling speed is limited. Finite element method (FEM) simulation has been carried out to study the deformation of thin oxide scale in hot strip rolling [40, 58, 59] and it can be verified by the results in [56–58]. The crack propagation in oxide scale has also been simulated [59] and it was found the roughness of the interface between the oxide scale and the steel substrate affects the crack propagation significantly due to its effect on materials flow. Rough interface results in large crack width remained after rolling.

It is not accurate to assume that the oxide scale grown on the low carbon steel to be perfectly adhering or fully brittle in hot rolling. Different oxide scale thickness and structures result in different crack width, crack spacing and extent of steel substrate flow through the gaps. The steel composition can significantly affect the state of the oxide scale. Kim et al. [60] investigated that the differences in failure of oxide scales formed on mild, Si–Mn, Mn-Mo and 316L stainless steel using a high-temperature tensile test under hot rolling conditions at entry to the roll gap. Through-thickness cracks and sliding were competitive mechanisms for oxide scale failure for mild steel, but oxide scales formed on Si–Mn, Mn-Mo steel within the temperature range of 783–1,200 °C failed only by through-thickness cracking. It was found [60] that the oxide scales formed on 316L at 1,074–1,200 °C failed only by through-thickness cracking. The 316L oxide scale formed at 1,200 °C for 800 s showed strong adherence to the steel surface even after 5 % strain.

Fletcher et al. [52] successfully developed FEM model by linking "macro"- and "micro"- scales of modeling to simulate the oxide scale behavior. The oxide scale was simulated as comprising numerous scale fragments joined together to form an oxide scale layer 10–100 μm thick [61]. Oxide scale failure including fracture, ductile behaviour and sliding was predicted by taking into account the stress-directed diffusion, fracture and adhesion of the oxide scale, strain, strain rate and temperature [6, 52, 61–63]. Considering the complexity of scale that

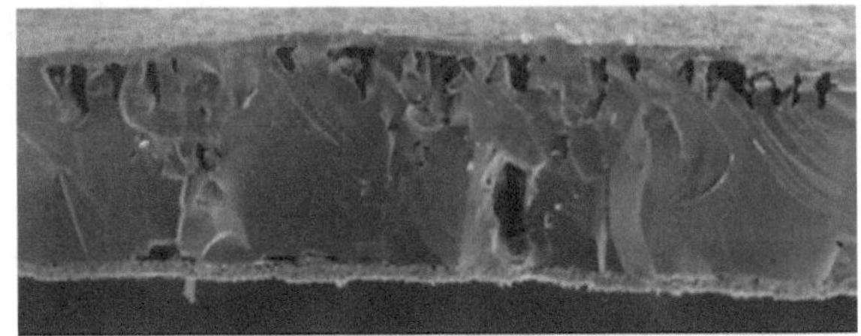

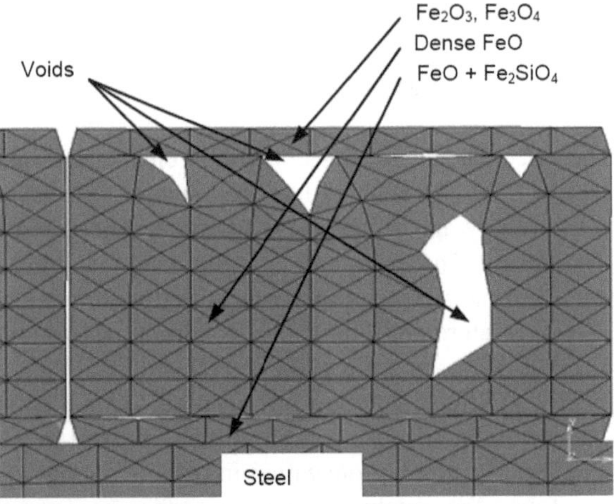

Fig. 4.12 SEM image (*top*) and FE mesh (*bottom*) representing the cross section of the three-layer steel oxide scale [63]

comprises 2–3 sub-layers, be porous, contains large voids and have a crystal being of the same order as the oxide scale thickness, the behaviour of multi-layer oxide scales could be simulated, as shown in Fig. 4.12. Numerical simulation on the behavior of the interface between the roll and the strip during multi-pass hot rolling operation was also carried out [64].

A primary result by FEM modelling [64] illustrated the possibility of simulation of main features of oxide scale failure behaviour on the deformation of the oxide scale on ferritic stainless steel during bending.

4.5.3.2 Stainless Steels

The situation of stainless steels is more complicated than that of plain carbon steels because oxide scales are not formed on the surface of stainless steels uniformly after reheating [65].

The microstructures of the oxide scale of 301 stainless steels before and after rolling when the reheating time is 25 and 35 min are illustrated in Figs. 4.13 and 4.14, respectively. The reheating temperature is 1,100 °C and entry temperature 1,050 °C. The oxide scale is thick and generally uniform. The outer oxide scale layers are detached from the inner oxide layer in the unrolled specimens due to the impact, and buoyancy resulted from the resin flow in the process of cold mounting.

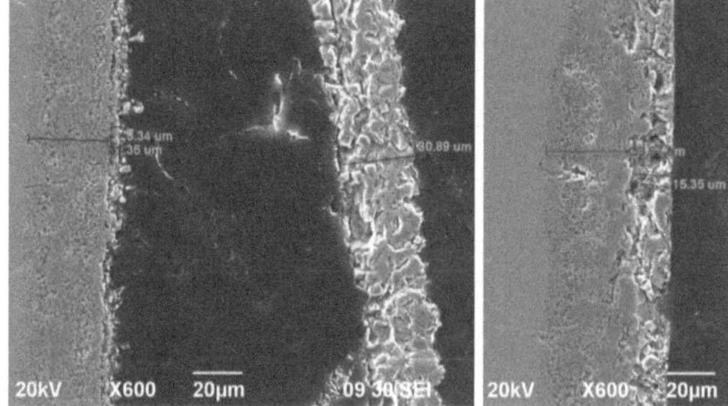

Fig. 4.13 Deformation of oxide scale of stainless 301 steel when reheating time is 25 min (*left*) before rolling, (*right*) after rolling at 19.9 % [65]

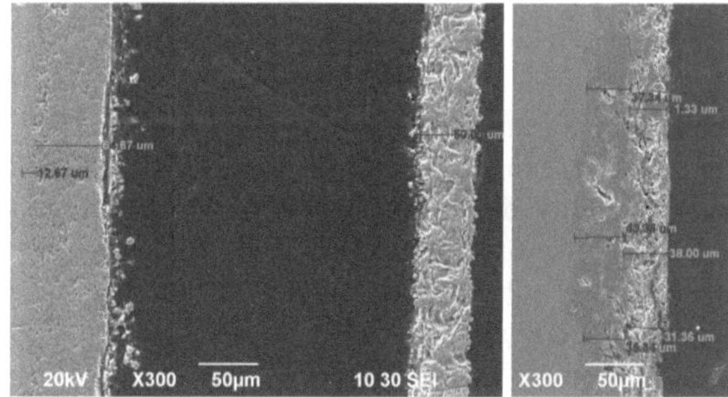

Fig. 4.14 Deformation of the oxide scale of stainless 301 steel when reheating time is 35 min (*left*) before rolling, (*right*) 20.1 % [65]

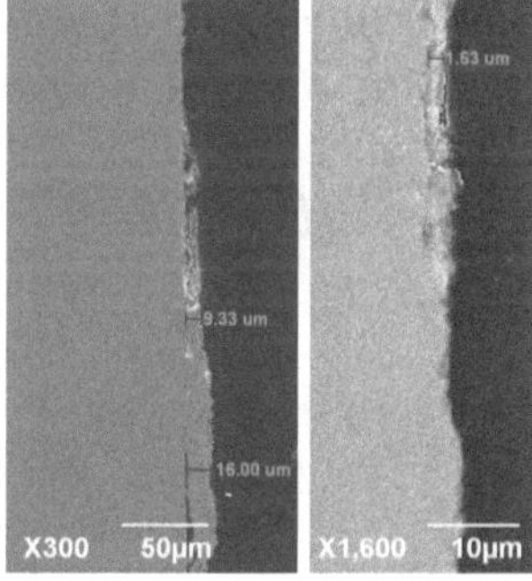

Fig. 4.15 Deformation of the oxide scale of stainless steel 430 when reheating time is 25 min (*left*) before rolling, (*right*) 28.3 % [65]

Both the number and size of the voids in the outer oxide scale layer are larger than that in the inner oxide layer.

The microstructures of the oxide scale of 430 stainless steels before and after rolling when the reheating time is 25 and 35 min are illustrated in Figs. 4.15 and 4.16, respectively. The reheating temperature is 1,050 °C and entry temperature 1,000 °C.

The oxidation of 430 stainless steel shows serious irregularity. The adhesion between the steel substrate and the oxide scale layer is high enough to resist the impact and buoyancy, resulting from the resin flow in the process of cold mounting.

The seriously non-uniform distribution of oxide scale results in very thin scale layer in some areas. The situation that the bare steel substrate contacts with the roll

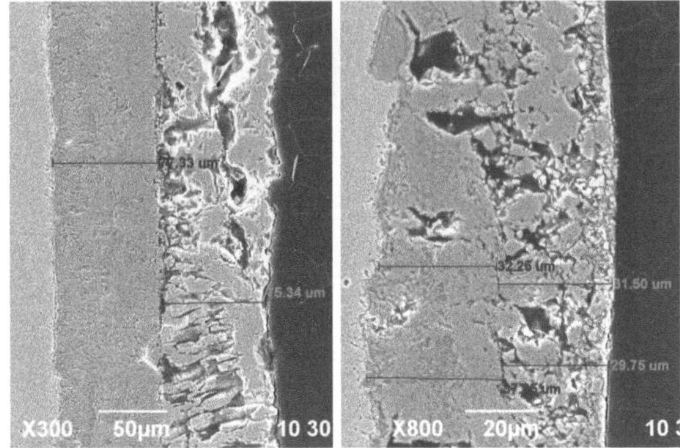

Fig. 4.16 Deformation of the oxide scale of stainless steel 430 when reheating time is 35 min (*left*) before rolling, (*right*) 21.9 % [79]

by extruding from the crack in oxide scale can happen in these areas, which results in sticking of ferritic stainless steel on the roll surface. Although there exists cracks normal to specimen surface in some areas where the oxide scale is thick, the compactability of the oxide scales is basically good, and the steel substrate is covered by the oxide scale very well, which may not result in the contact between the bare steel substrate and the roll. So sticking of steel on the roll surface cannot happen when the oxide scale is thick.

4.6 Oxide Scale Involved Surface Roughness and its Transfer During Hot Rolling

4.6.1 Carbon Steel

Oxide scale of carbon steel may be considered either thin and hot or thick and cold when it enters the finishing mill [11]. In the first case, the scale fractures along fine lines in the first stand, and hot metal is extruded through these fine lines, which results in a simultaneous roughening and smoothing of the steel surface. In the second case, the oxide scale is less plastic and fractures severely during deformation. It occurs at the same time that the fractured oxide scale is depressed into the steel surface and hot metal extrudes outwards, which results in a rough surface.

It was found that the surface roughness decreases remarkably in hot rolling test of low carbon steel when the bulk deformation increases in the case of thick primary oxide scale, but the ability to decrease surface roughness by the deformation of thin secondary oxide scale is very limited [56]. In the case of 55–70 min of heating, the roughness of both the descaled and non-descaled increases with rolling speed, while the surface roughness decreases with increasing rolling speed when the sample heating time is 145-160 min. Heavy reduction has a significant influence on the surface finish, as shown in Fig. 4.17.

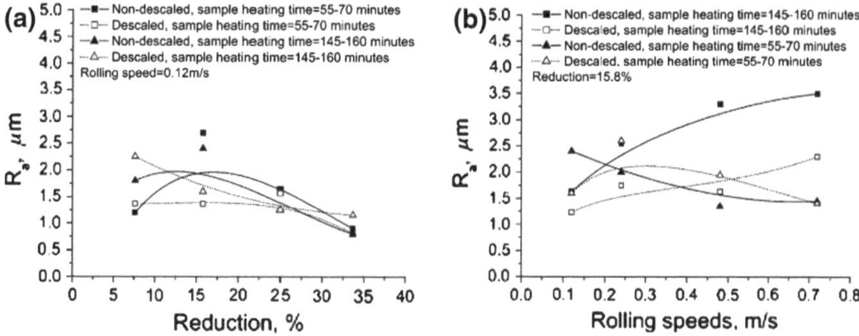

Fig. 4.17 Effects of reductions and rolling speeds on surface roughness [56]

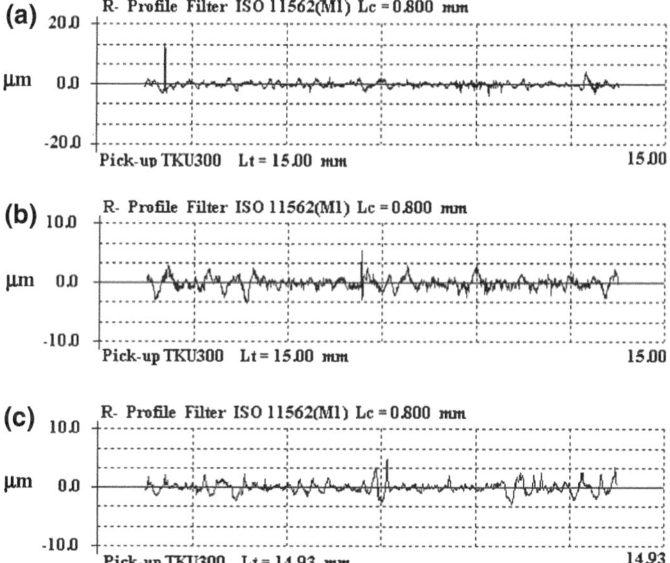

Fig. 4.18 Surface profiles of 304L after hot rolling under different rolling reduction when reheating time is 25 min. Reduction is **a** 10.6 %, **b** 20.6 %, **c** 29.3 %

4.6.2 Stainless Steels

The rough outmost surface of oxide scale of stainless steels before rolling becomes smooth after rolling [65]. The surface status does not keep constant at all positions on each sample because the distribution of oxide scale on the surface is not very uniform. It was found that some surface area was relatively rougher than other area especially when the reduction was 10 and 20 %, and the rougher area usually appeared dark red. This may be because the outmost Fe_2O_3 oxide scale layer does not experience enough compression, and still break up during cooling. In order to

Table 4.2 Surface roughness of 304L after hot rolling under different reduction

Reheating time (min)	Reduction (%)	R_a, (μm)	R_{max}, (μm)
	10.6	0.83	16.66
25	20.6	0.72	8.29
	29.3	0.65	7.39
	10.6	1.05	22.77
35	20.1	1.00	17.55
	29.4	0.75	6.54

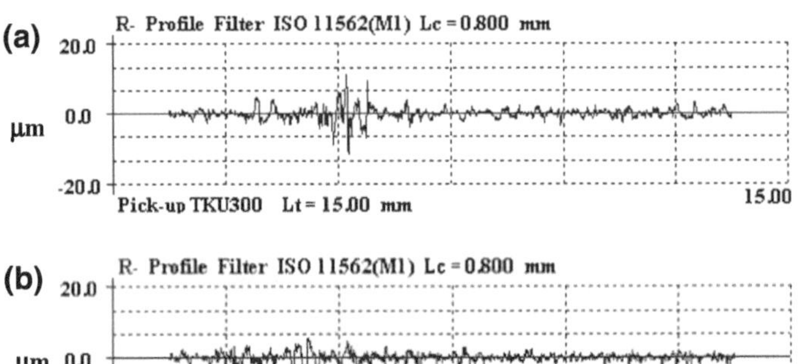

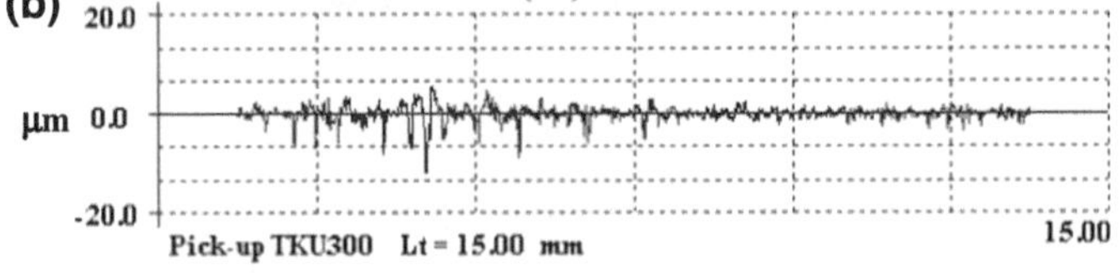

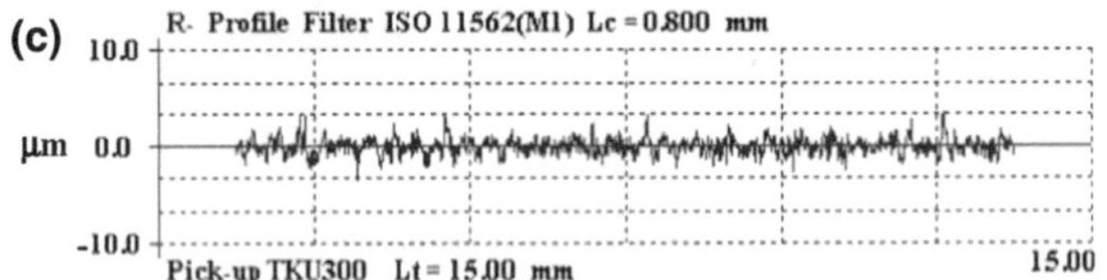

Fig. 4.19 Surface profiles of 304L after hot rolling under different rolling reduction when reheating time is 35 min. Reduction is **a** 10.6 %, **b** 20.1 %, **c** 29.4 %

compare the surface status under different reductions, the relatively smooth surface area on each sample was measured in all cases. However, the surface roughness measurement is more qualitative than quantitative.

When reheating time is 25 and 35 min, the surface profiles of stainless steel 304L along rolling direction under various reductions are shown in Figs. 4.18 and 4.19 respectively.

The surface roughness parameters R_a and R_{max} of 304L after hot rolling under different reductions are shown in Table 4.2 [66].

Generally, the values of R_a and R_{max} in the case of reheating time 35 min are larger than that in the case of reheating time 25 min, except that the value of R_{max} in the case of reheating time 35 min is a little bit smaller than that in the case of reheating time 25 min when the reduction is nearly 30 %.

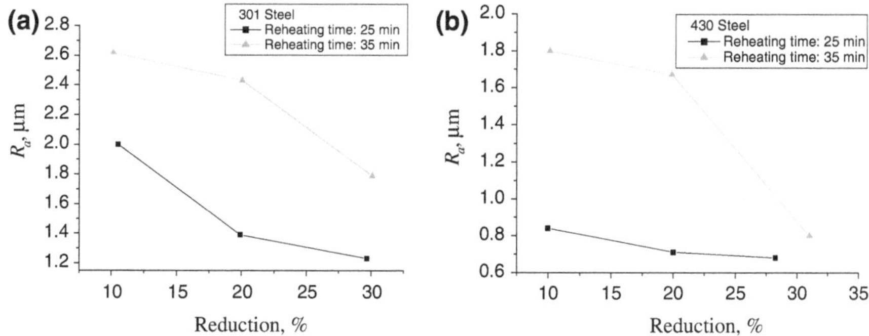

Fig. 4.20 Relationship between R_a and rolling reduction [65]

R_a and R_{max} always decrease with an increase of rolling reduction. In the case of 25 min reheating time, R_a decreases 13.3 % and R_{max} decreases 50.2 % when the reduction increases from 10 to 20 %, while they decrease 9.7 and 10.9 % respectively when reduction increases from 20 to 30 %. This means that the surface roughness decreases more dramatically when the reduction is lower than 20 % than beyond 20 %. On the contrary, in the case of 35 min reheating time, R_a decreases 4.8 % and R_{max} decreases 22.9 % when the reduction increases from 10 to 20 %, while they decrease 25.0 and 62.7 % respectively when reduction increases from 20 to 30 %, which shows that the surface roughness decrease more significantly when the rolling reduction is beyond 20 % than lower than 20 %.

As mentioned above, the thickness of oxide scale is 34 and 80 μm respectively when the reheating time is 25 and 35 min. On the one hand, the oxide scale layer is not compact when the rolling reduction reaches 20 % [10] so thicker oxide scale results in relatively rougher 304L strip surface during hot rolling. When the rolling reduction reaches 30 %, the oxide scale layer tends to be compact [10], which results in very near surface roughness in both cases. On the other hand, when the thickness of oxide scale is 34 μm, the reduction of 20 % is enough to decrease the surface roughness significantly. However, when the thickness of oxide scale is 80 μm, the reduction should be 30 % to achieve the significant decrease of surface roughness.

The relationship between R_a of 301 and 430 steels along rolling direction and reduction is shown in Fig. 4.20. Generally, the values of R_a decrease with an increase of reduction. This shows that the height of the surface asperities of oxide scale decreases with an increase of reduction. On the other hand, the compactability of oxide scale increases with an increase of reduction, which can help reduce the breaking in the oxide scale layer, which also has a contribution to the decrease of surface roughness.

R_a when the reheating time is 35 min is larger than that when the reheating time is 25 min at all reductions. R_a of 301 stainless steel is higher than that of 430 stainless steel when the reheating time is 25 and 35 min, respectively. In the case of 301 stainless steel, as the thickness of oxide scale when the reheating time is

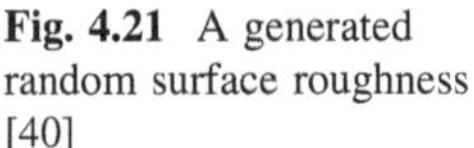

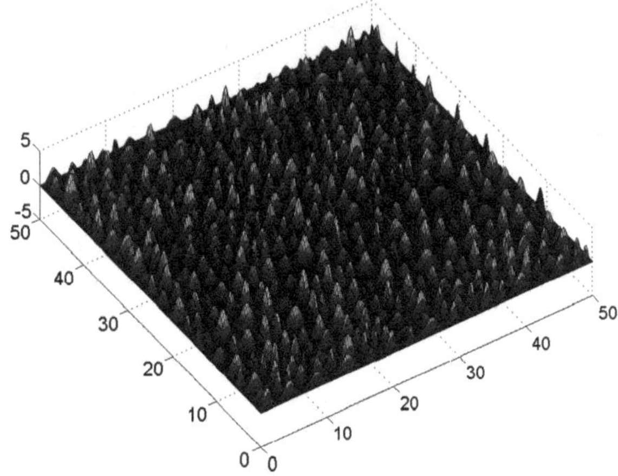

Fig. 4.21 A generated random surface roughness [40]

35 min is larger than that when the reheating time is 25 min, and the thick outer layer owns relatively low compactability and is prone to be broken, which results in relatively rough strip surface after hot rolling. In the case of 430 stainless steel, the breaking of oxide scale is not significant and the seriously non-uniform distribution of oxide scale as described above should be responsible for this result.

4.6.3 Numerical Simulation

Tang et al. [40, 58, 59] studied the thin secondary oxide scale deformation and surface roughness transfer in hot rolling. Profile of roughness asperity was described by a normal function.

$$z = A \cdot e^{-[(x-\mu)^2/\sigma_x^2+(y-v)^2/\sigma_y^2]/2} \quad (4.3)$$

where z and A are the height and the peak height of the roughness asperity respectively, μ and v are the summit coordinates of the roughness asperity respectively, σ_x and σ_y are standard deviations that reflect the wavelength or sharpness of the roughness asperity.

Based on Eq. (4.3), the generated rough surface is shown in Fig. 4.21, which has the mathematically characteristics of the surface of real oxide scale of plain carbon steel shown in Fig. 4.3 and that of stainless steel shown in Fig. 4.22, was adopted for FEM simulation. A finite element model for rolling process including the oxide scale layer and surface roughness with meshing was developed as shown in Fig. 4.23.

The simulation results show that the final surface roughness of steel substrate is always larger than that of oxide scale and both increase with the initial roughness. The oxide scale thickness has a significant effect on the final surface roughness of steel and thick oxide scale layer results in large final surface roughness. The surface roughness of work rolls has a significant effect on the surface roughness transfer in

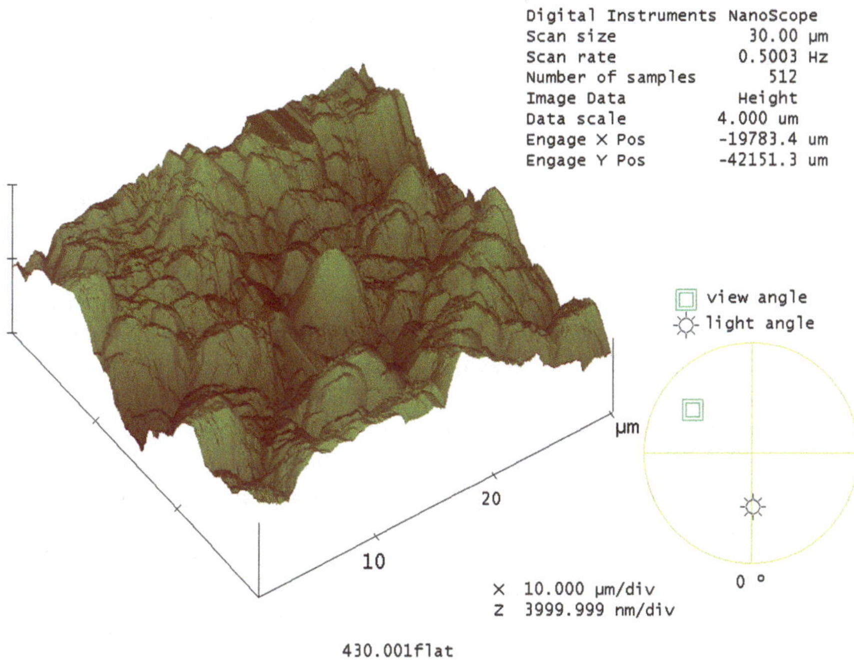

Fig. 4.22 Oxidized 430 stainless steel strip surface by Atomic Force Microscope

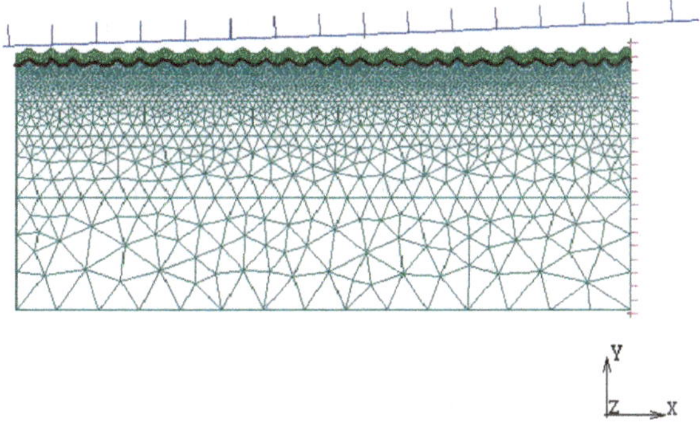

Fig. 4.23 FE model for rolling process with oxide scale layer and surface roughness [40]

hot rolling. In the case of the surface roughness of work roll less than 1 μm, the surface roughness of thin oxide scale and strip decreases significantly during hot rolling. However, if the surface roughness of work roll is larger than 1 μm, the decrease of the final surface roughness of the strip becomes small at best.

Two methods were suggested to improve the surface quality of hot rolled steel, maintaining continuous and uncracked oxide scale on the metal surface during rolling process or completely removing oxide scale before stock enters roll gap. Based on the latter, the simulation of descaling has been carried out to optimise the parameters [64].

4.7 Evaluation of Friction in Hot Rolling and Tribological Effect of Oxide Scale

4.7.1 Evaluation of Friction in Hot Rolling

Friction is one of the most significant physical phenomena influencing the forming of metals but it remains the least understood in comparison with metallurgy, heat transfer and mechanics [67]. A large number of tests have been conducted to measure friction, however, very few are suitable for hot metal forming application [67]. A detailed study of friction is important for a thorough understanding of the mechanical and metallurgical behaviour of hot rolling steels. Both direct measurement and indirect methods have been developed to determine friction.

The measurement of friction is difficult due to its variation in roll gap. The direct measurement method such as inscribed gridlines and embedded pin inserts has been developed. The inscribed gridlines is more expensive and time consuming than the embedded pin inserts techniques [68]. Embedded pin transducer systems can provide much useful information on friction in hot strip rolling [68–70], but it is very difficult to manage and operate. It does not fit for hot rolling where the oxide scale forms on the steel surface [71]. The accuracy strongly depends on the positions of the pin transducers relative to the surface of tool [71].

Indirect estimation methods have been adopted because of the above limitation. Lundberg [71] adopted forward slip measurements under normal production conditions to evaluate the coefficient of friction in the hot rolling of steel bars. Considering the effect of oxide scale, Li et al. [72] developed forward/backward slip measurement method for determining friction in hot strip rolling. In this method, the neutral angle was derived from forward and backward slip values. Assuming an Amontons-coulomb friction law, the constant roll pressure along the arc contact and constant slab width, an equation has been established for determining the relationship between the neutral angle and the friction coefficient. Other methods include bite angle measurements, ring upsetting tests etc.

4.7.2 Tribological Effect of Oxide Scale

The deformation or fracture of oxide scale affects the interface between the roll and the strip due to (1) the thermal conductivity of oxide scale is much less than that of steel substrate; (2) the crack in scale enables hot substrate to contact with the cold roll, and (3) possibility of sliding of the fractured scale raft on substrate. The oxide scale on hot steel substrate is the key for the control of friction and heat transfer at the interface between the roll and the strip [12].

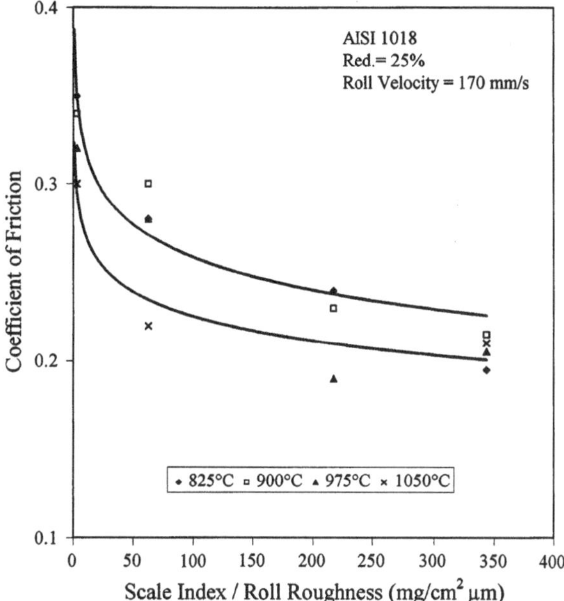

Fig. 4.24 The coefficient of friction as a function of oxide scale index and roll roughness (scale index expressed in mass gain per unit surface area)

4.7.2.1 Carbon Steels

Munther et al. [11] investigated oxide scale formation and coefficient of friction in the hot flat rolling of steels. It was concluded that the oxide scale thickness affects the coefficient of friction with temperature change. A thick oxide scale provides lubricity and generates a lower coefficient of friction while a thin oxide scale is more adherent to the metal substrate and related to a higher coefficient of friction, as shown in Fig. 4.24.

Any material between two contacting surfaces should be considered as a lubricant [73]. The normal roughness of the roll surface and degree of oxide scale can account for the force separating the roll to a magnitude of 44 % and for torque up to 58 % [74].

Yu and Lenard [9] investigated the influence of oxide scale on the coefficient of friction in hot steel rolling of carbon steel strip. It was pointed out that the most significant parameter determining the coefficient of friction is the thickness of oxide scale layer in between the rolled stock and the work roll. Generally, thicker oxide scales cause lower friction. Fedorciuc-Onisa et al. [75] took into consideration the complex interactions at the stock-roll interface due to the presence of secondary oxide scale, and proposed a new Coulomb-Newton type friction model for long products and bar sections.

If oxide scale is thin, it may not act as a lubricant because it adheres to the substrate [11, 55]. Luong and Heijkoop [2] investigated the influence of both oxide scale thickness and oxide scale composition on interfacial friction in hot metal working. A number of factors, including porosity, oxides present in the layer, and

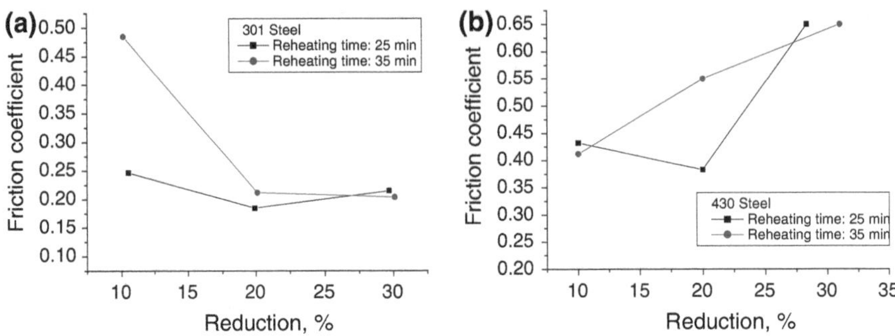

Fig. 4.25 Relationship between friction coefficient and rolling reduction during hot rolling of 301 and 430 steels [65]

the chemical composition of the steel influence the formation of wustite, determine the strength of the oxides and therefore the coefficient of friction.

On the other hand, oxide scale can help reduce wear by protecting against loss of metal due to mechanical damage [7]. It is also beneficial for the tool surface when it acts as a lubricant in hot metal forming, but it may become abrasive if it comes off the work-piece and cause damage to the tool surface [2]. Vergne et al. [76] investigated the evolution of the friction coefficient and the wear on the surface of hot working tool using a high temperature pin on disc tribometer. The results show that the oxide scales play a lubricating role in friction.

4.7.2.2 Stainless Steels

A calculation code developed by Alexander on the basis of Orowan's model [77] was adopted for inversely determining the coefficient of friction in the processes of hot rolling of stainless steels [78]. It is capable of handling the mixed boundary condition at the roll-material interface. The coefficients of friction were varied until the calculated rolling force is less than 1 % error with the measured one in each case.

The relationship between the reduction and calculated coefficient of friction during hot rolling is shown in Fig. 4.25. In the case of 301 stainless steel, the calculated values of coefficients of friction decrease with an increase of rolling reduction. When the reheating time is 35 min, which results in thick oxide scale, the difference of the values of coefficients of friction between the reduction of 10 and 20 % is much larger than that between 20 and 30 %. This means that the thicker oxide scale shows higher lubricative effect, and the microstructure of oxide scale of 301 stainless steel becomes stable after hot rolling at the reduction of 20 %, so the lubricative effect also becomes stable.

On the contrary, the calculated values of coefficients of friction basically increase with an increase of reduction in the case of 430 stainless steel. The values reach 0.65 when the reduction is around 30 %, which is quite high. This may also result from the non-uniform oxide scale no matter what the reheating time is 25 or 35 min. The thin oxide scales break on many places, which make the steel substrate to extrude from crack to contact the roll. This counts the lubricative effect of oxide scale, and increases the coefficient of friction. The number of breakings of oxide scales increases with an increase of reduction, so the values of coefficients of friction increase with an increase of reduction. Sticking may occur under this situation. If a relative thick and uniform oxide scale layer can be generated to reduce the breaking of the oxide scale, not only will the friction condition be improved, but also the problem of sticking in the hot rolling of ferritic stainless steel can be alleviated.

References

1. Torresa M, Colas R (2000) J Mater Process Technol 105:258–263
2. Luong LH, Heijkoop T (1981) Wear 71:93–102
3. Li YH, Krzyzanowski M, Beynon JH, Sellars CM (2000) Acta Metall Sin 13:359–368
4. Beynon JH, Li YH, Krzyzanowski M, Sellars CM (2000) In: Pietrzk, Kusiak J, Majta J, Hartley P and Pillinger I (eds) Metal forming, Balkema, Rotterma, pp 3–10
5. Krzyzanowski M, Beynon JH (2000) Modell Simul Mater Sci Eng 8:927–945
6. Krzyzanowski M, Beynon JH, Sellars CM (2000) Metall Mater Trans 31B:1483–1490
7. Krzyzanowski M, Beynon JH (1999) Steel Res 70:22–27
8. Chen RY, Yuen WYD, Mak T (2001) In Proceedings of the 43rd MWSP conference, ISS, vol XXXIX, pp 287–299
9. Yu Y, Lenard JG (2002) J Mater Process Technol 121:60–68
10. Li YH, Sellars CM (1996) In: Beynon JH, Ingham P, Teichert H and Waterson K (eds) Proceedings of the 2nd international conference on modelling of metal rolling processes, London, pp 192–206
11. Munther PA, Lenard JG (1999) J Mater Process Technol 88:105–113
12. Krzyzanowski M, Beynon JH (2006) ISIJ Int 46:1533–1547
13. Chen RY, Yuen WYD (2003) Oxid Met 59:433–468
14. Ajersch F (1993) In Proceedings of the 34th MWSP conference, ISS-AIME 30, pp 419–437
15. Paidassi J (1958) Acta Metall 6:184–194
16. Bertrand N, Desgranges C, Gauvain D, Monceau D, Poquillon D (2004) Mater Sci Forum 461–464:591–598
17. Singh Raman RK, Gleeson, Young DJ (1996). In Proceedings of the 13th international conference on corrosion, Australia, paper 297
18. Abuluwefa H, Guthrie RIL, Ajersch F (1996) Oxid Met 46:423
19. Chen RY, Yuen WYD (2002) Oxid Met 57:53–79
20. Evans HE (1988) Mater Sci Technol 4:1089
21. Mark R (2001) Wire industry, 68, p 503
22. Capitan MJ, Lefebvre S, Traverse A, Paul A, Odriozolac JA (1998) J Mater Chem 8:2293–2298
23. Riffard F, Buscail H, Caudron E, Cueff R, Issartel C, Perrier S (2004) Surf Eng 20:440–446
24. Riffard F, Buscail H, Caudron E, Cueff R, Issartel C, Perrier S (2002) Mater Charact 49:55–65
25. Riffard F, Buscail H, Caudron E, Cueff R (2001) In: Sudarshan TS, Jeandin M (eds) Surface modification technologies XIV, ASM international, materials park. Ohio and IOM Communications Ltd., UK, pp 526–529
26. Jonsson T, Canovic S, Liu F, Asteman H, Svensson J-E, Johansson L-G, Halvarsson M (2005) Materials at high temperature 22:231–243

27. Asteman H, Svensson J-E, Johansson L-G (2002) Oxid Met 57:193–216
28. Asteman H, Svensson J-E, Johansson L-G, Norell M (1999) Oxid Met 52:95–111
29. Asteman H, Svensson J-E, Norell M, Johansson L-G (2000) Oxid Met 54:11–26
30. Asteman H, Segerdahl K, Svensson JE, Johansson LG (2001) Mater Sci Forum 369–372:277–286
31. Tang JE (2001) Micron 32:799–805
32. Honda Katsuya, Maruyama Toshio, Atake Tooru, Saito Yasutoshi (1992) Oxid Met 38:347–363
33. Ishida Toshihisa, Harayama Yasuo, Yaguchi Sinnosuke (1986) J Nucl Mater 140:74–84
34. Cheng Shen-Yuan, Kuan Sheng-Lih, Tsai Wen-Ta (2006) Corros Sci 48:634–649
35. Amy S, Vangeli P (2001) In: EUROCORR 2001 (The European Corrosion Congress), Lake Garda, Italy, 2001, 13-25
36. Huntz AM, Reckmanna A, Haut C, Severac C, Herbst M, Resende FCT, Sabioni ACS (2007) Mater Sci Eng, A 447:266–276
37. Lee YD, Lee YH, Lee JS, Kim JK (1991) In Proceedings of the international conference on stainless steels. Chiba, ISIJ, pp 952–958
38. Sun WH, Tieu AK, Jiang ZY, Zhu H, Lu C (2003) J Mater Process Technol 140:76–83
39. Sun WH, Tieu AK, Jiang ZY, Zhu H, Lu C (2004) J Mater Process Technol 155–156:1300–1306
40. Jiang ZY, Tieu AK, Sun WH, Tang JN, Wei DB (2006) Mater Sci Eng A 435–436:434–438
41. Perez FJ, Cristobal MJ, Arnau G, Hierro MP, Saura JJ (2001) Oxid Met 55:105–118
42. Echsler H, Ito S, Schutze M (2003) Oxid Met 60:241–269
43. Chen RY, Yuen WYD (2000) Oxid Met 53:539–560
44. Chen RY, Yuen WYD (2001) Oxid Met 56:89–118
45. Fernando LA, Zaremski DR (1988) Metall Trans A 19:1083–1100
46. Lundberg S-E, Gustafsson T (1994) J Mater Process Technol 42:239–291
47. Krzyzanowski M, Beynon JH (2002) J Mater Process Technol 125–126:398–404
48. Hidaka Y, Anraku T, Otsuka N (2001) Mater Sci Forum 369–372:555–562
49. Riedel H (1982) Mater Sci 16:569–574
50. Hancook P, Nicholls JR (1988) Mater Sci Technol 4:398–406
51. Schutze M (2005) Materials at high temperature 22:147–154
52. Fletcher JD, Beynon JH (1996) In Proceedings of the second international conference on modelling of metal rolling processes, London, 202–212
53. Li YH, Sellars CM (1996) Ironmaking steelmaking 23:58–61
54. Li YH, Sellars CM (1999) Advanced technology of plasticity. In Proceedings of the 6th ICTP, vol III, 1973–1978
55. Jarl M (1993) In Proceedings of the first international conference on modelling of metal rolling processes. UK, London, pp 614–628
56. Sun WH, Tieu AK, Jiang ZY, Lu C (2004) J Mater Process Technol 155–156:1307–1312
57. Sun WH, Tieu AK, Jiang ZY, Zhu H (2004) Key Eng Mater 274–276:511–516
58. Tang J, Tieu AK, Jiang ZY (2006) J Mater Process Technol 177:126–129
59. Tang J, Tieu AK, Jiang ZY (2004) Key Eng Mater 274–276:499–504
60. Tan KS, Krzyzanowski M, Beynon JH (2001) Steel Res 72:250–257
61. Krzyzanowski M, Beynon JH (1999) Mater Sci Technol 15:1191–1198
62. Krzyzanowski M, Beynon JH (1999) In Proceedings of the third international conference on modelling of metal rolling processes, London, 360–369
63. Beynon JH, Krzyzanowski M (2005) Mater Forum 29:39–46
64. Krzyzanowsky M, Beynon JH, Frolish MF, Clowe S (2007) Mater Sci Forum 539–543:2461–2466
65. Wei DB, Huang JX, Zhang AW, Jiang ZY, Tieu AK, Shi X, Jiao SH (2010) Friction, surface roughness and oxide scale deformation during hot rolling of stainless steels. In proceedings of the 10th international conference on steel rolling, Beijing, China, 15–17
66. Wei DB, Huang JX, Zhang AW, Jiang ZY, Tieu AK, Shi X, Jiao SH (2009) Int J Surf Sci Eng 3(5–6):459–470
67. Rudkins NT, Hartley P, Pillinger I, Petty D (1996) J Mater Process Technol 60:349–353
68. Das S, Palmiere EJ, Howard IC (2001) Mater Sci Technol 17:864–873

69. Hum B, Colquhoun HW, Lenard JG (1996) J Mater Process Technol 60:331–338
70. Lagergren J (1997) J Mater Process Technol 70:207–214
71. Lundberg S-E (2004) Scandinavian J Metall 33:129-145
72. Li YH, Beynon JH, Sellars CM (1999) Advanced technology of plasticity. In Proceedings of the 6th ICTP, vol III, 2023–2028
73. Heshmat H, Godet M, Berthier Y (1995) Lubr Eng 51:557–564
74. A.K.E.H.A. El-Kalay, Sparling LGM (1968) J Iron and Steel Inst 43:152-168
75. Fedorciuc-Onisa C, Farrugia DCJ (2003) In: Brucato V (eds) The 6th ESAFORM conference on material forming, 763–766
76. Vergne C, Boher C, Levaillant C, Gras R (2001) Wear 250:322–333
77. Alexander JM, Brewer RC, Rowe GW (1987) Manufacturing technology, Vol 2: engineering processes. Ellis Horwood Ltd. Published, West Sussex
78. Wei DB, Huang JX, Zhang AW, Jiang ZY, Tieu AK, Wu F, Shi X, Jiao SH, Chen L (2010) Steel Res Int 81(9):102–105
79. Wei DB, Huang JX, Zhang AW, Jiang ZY, Tieu AK, Shi X, Jiao SH (2011) Wear 271(9–10):2417–2425

Chapter 5
Micro-Contact at Interface Between Tool and Workpiece in Metalforming

Akira Azushima

Metalforming tribology is a science and technology of a huge system. At the midst of this system is situated the micro-contact mechanism. The micro-contact mechanism in the presence of lubricant have generally been classified into the regimes of boundary, mixed and hydrodynamic lubrication depending on the average lubricant film thickness and the surface roughnesses of tool and workpiece. In these mechanism, there proceeds mechanical interaction between the surface microstructures of workpiece and tool and the lubricant. First, this paper presents the micro-contact at the interface between tool and workpiece under each lubrication regime. Second, the micro-contact at the interface between tool and workpiece in sheet metals forming, and sheet rolling are explained.

5.1 Introduction

Knowledge on the interaction between surface asperities and lubricant in the micro-contact mechanism is of great important for the understanding of tribological effects and the improvement of tribological techniques in cold metal forming processes [1]. The micro-contact mechanism in the presence of lubricant have generally been classified into the regimes of boundary, mixed and hydrodynamic lubrication depending on the lubricant film thickness h and the surface roughnesses of solid 1 and solid 2 as shown in Fig. 5.1. This figure is called “Stribeck Diagram”.

A. Azushima (✉)
Department of Mechanical Engineering, Yokohama National University,
79-5 Tokiwadai Hodogaya-ku Yokohama 204-8501, Japan
e-mail: azu@ynu.ac.jp

J. P. Davim (ed.), *Tribology in Manufacturing Technology*, Materials Forming, Machining and Tribology, DOI: 10.1007/978-3-642-31683-8_5,

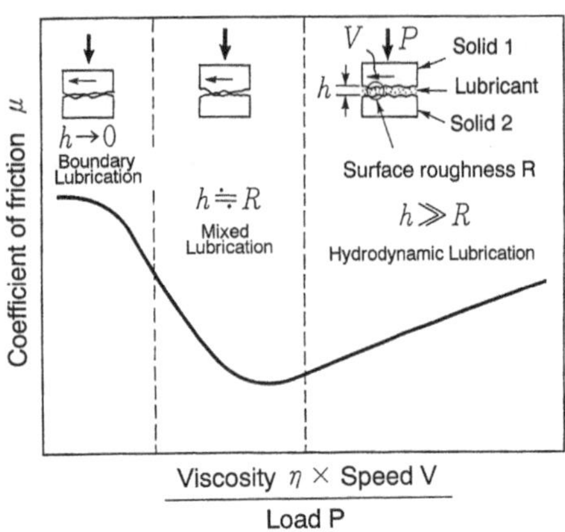

Fig. 5.1 Stribeck diagram

In Fig. 5.1, the λ value is recently used in order to classify quantitatively the lubrication regime. The λ value is expressed by the Eq. (5.1).

$$\lambda = \frac{h_a}{\sigma} \tag{5.1}$$

where, h_a is the average oil film thickness and σ is the combined surface roughness. When λ is larger than 4, the lubrication regime is "Macro-Plasto-Hydrodynamic Lubrication". Next. When λ is larger than 1 and smaller than 4, the regime is "Mixed lubrication". When λ is smaller than 1, the regime is "Boundary Lubrication. In this paper, the micro-contact at the interface between tool and workpiece under each lubrication regime are reviewed.

5.2 Micro-Contact Under Each Lubrication

5.2.1 Micro-Contact Under Dry and Thin Film Lubrication

The plastic flattening of asperities of workpiece surface by flat tool surface in the absence of the macroscopic sliding is explained. A concept of the real contact area A_{r0} which corresponds to the flattened area of asperities is used as shown in Fig. 5.2.

The magnitude of real contact area is given by the Eq. (5.2),

$$A_{ro} = \frac{P}{p_m} \approx \frac{P}{HV} \tag{5.2}$$

where P denotes the load given perpendicularly to the apparent contact interface and p_m denotes the pressure prevailing over the real contact area and pm equals the value of the indentation.

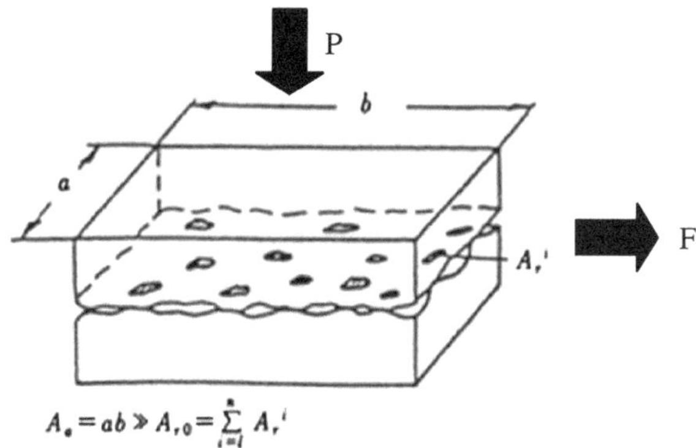

Fig. 5.2 Micro-contact model

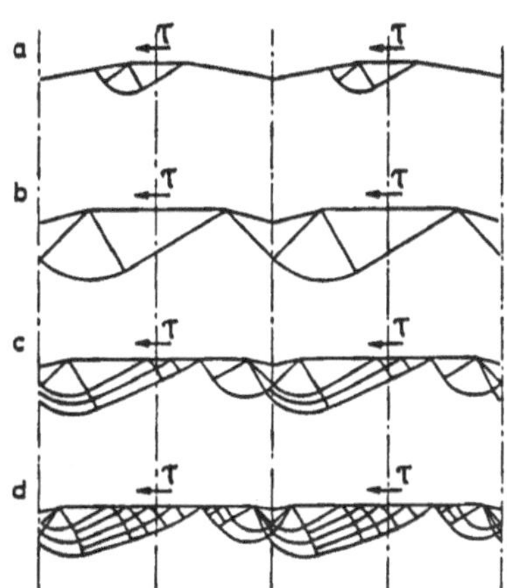

Fig. 5.3 Slip line field of regular arrays of wedge-shape ridges deforming [3]

When the tangential force F in the presence of the macroscopic sliding acts and the coefficient of friction is μ, the real contact area A_r increases as shown in the Eq. (5.3) proposed by MacFarlane and Tabor [2].

$$A_r = A_{ro}\sqrt{1 + \alpha\mu^2} \tag{5.3}$$

On the other head, the slip-line solution for regular arrays of wedge-shape ridges deforming under a high pressure and with an interfacial shear stress in combination with the asperities is reported by Bay and Wanheim [3]. Figure 5.3 shows the slip line field and Fig. 5.4 shows the relationship between the real contact area ratio and the non- dimensional pressure.

The theoretical results show that the value of the shear factor m markedly influences the asperity deformation. When the m value becomes higher, the real contact area ratio increases rapidly with growing the non-dimensional pressure. It is anticipated that the sliding can influence the m value by generating heat or by damaging the boundary film. In the observation so far made above the underling bulk of the workpiece is assumed rigid. It is well known that the real contact area ratio increases by the bulk deformation.

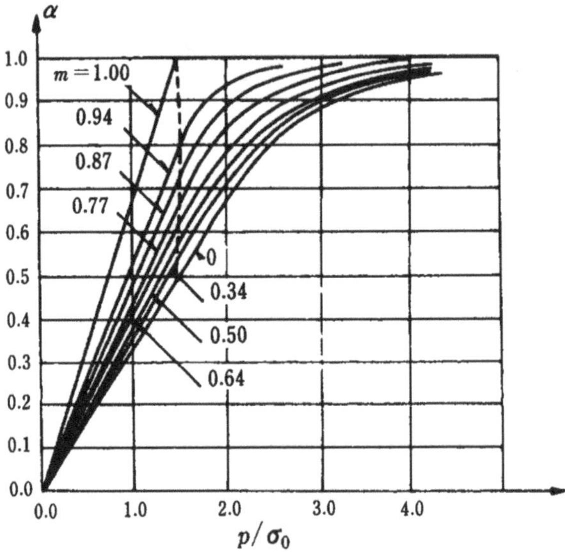

Fig. 5.4 Relationship between real contact area ratio and non-dimensional pressure [3]

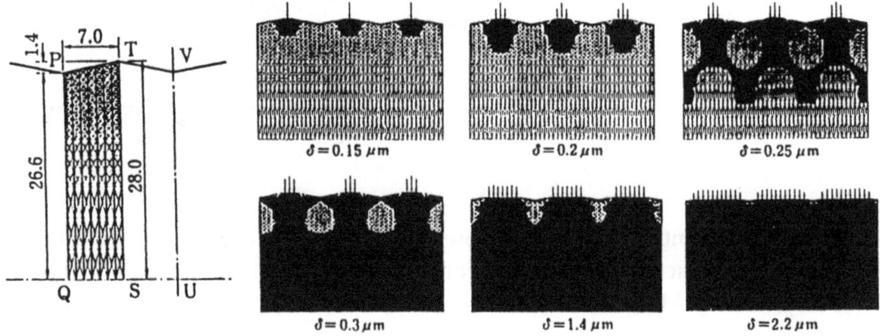

Fig. 5.5 Elastic-plastic finite element analysis of regular array of wedges [4]

Makinouci et al. [4] used an elastic–plastic finite element method to analyze the plane-strain upsetting of a specimen having a regular array of wedges as shown in Fig. 5.5. The result as shown in Fig. 5.6 indicates that the real contact area becomes nearly unity already at a workpiece reduction in height of 9 %. In these analyses, the friction at the real contact area is ignored.

In this lubrication mechanism, the sheet with a mirror surface is produced by skin pass rolling. However, in the rolling process the productivity with higher rolling speed can not be raised due to the occurrence of friction pick up.

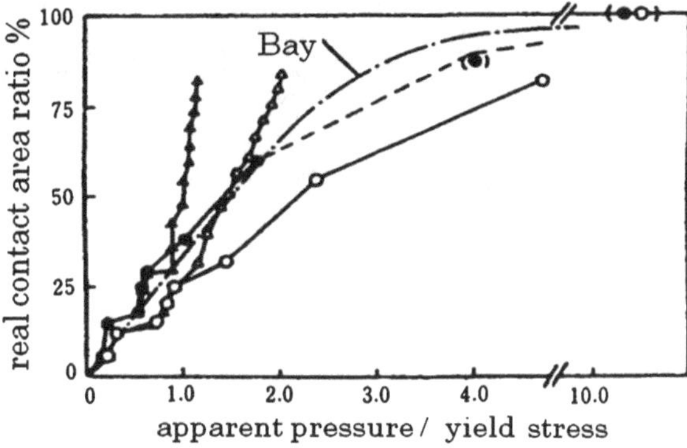

Fig. 5.6 Relationship between real contact area ratio and non-dimensional pressure [4]

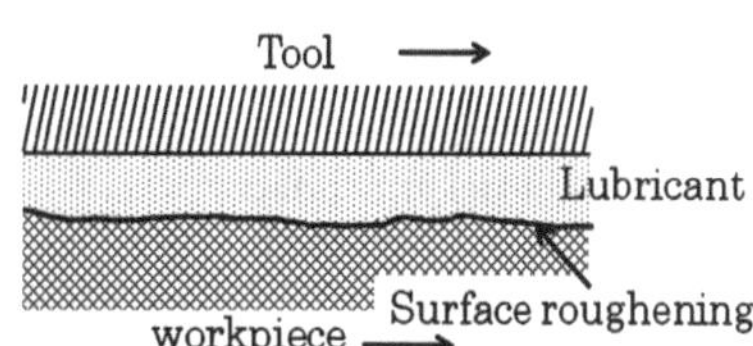

Fig. 5.7 Model of macro-plasto-hydrodynamic lubrication

5.2.2 Micro-Contact Under Thick Film Lubrication

Figure 5.7 shows the model of the macro-plasto-hydrodynamic lubrication. The thick oil film exists at the interface between tool and workpiece. The oil film thickness at the entrance in steady state metal forming is calculated based on the well-known Reynolds equation, with the assumptions that the lubricant undergoes steady incompressible flow and that both smooth surface of tool and workpiece and do not deform outside the working area [5]. At the beginning of plastic deformation, the thick film is trapped between tool and workpiece and the micro-contact does not occur between surfaces of tool and workpiece. As the plastic deformation proceeds, the free surface roughening on the workpiece surface. As the free surface roughening becomes larger, asperities on the workpiece surface start to contact.

The surface roughness of the workpiece surface deformed freely is given by this Eq. (5.4).

$$R_{\varepsilon}^{Mate} = R_0 + C(\varepsilon - \varepsilon_0)D \tag{5.4}$$

The surface appearance of workpiece after rolling under the macro-plasto-hydrodynamic lubrication is shown in Fig. 5.8a. The appearance is similar to the surface appearance of workpiece after tensile test as you can see from Fig. 5.8b.

Then, the shear stress is expressed as the Eq. (5.5).

$$\tau = \frac{\partial u}{\partial y} \tag{5.5}$$

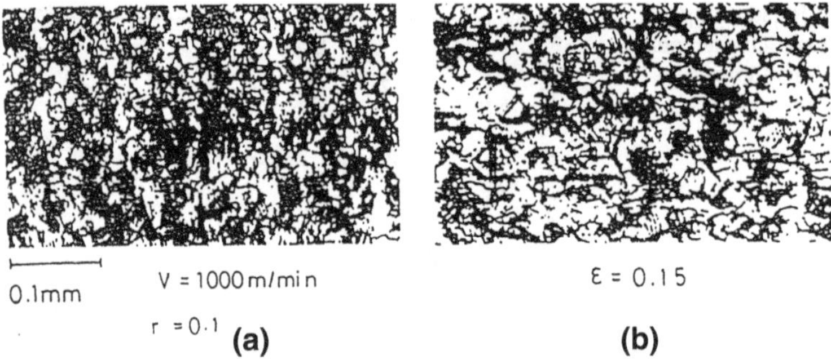

Fig. 5.8 Surface appearance of workpiece after rolling under thick film lubrication (**a**) and surface appearances of workpiece after tensile test (**b**)

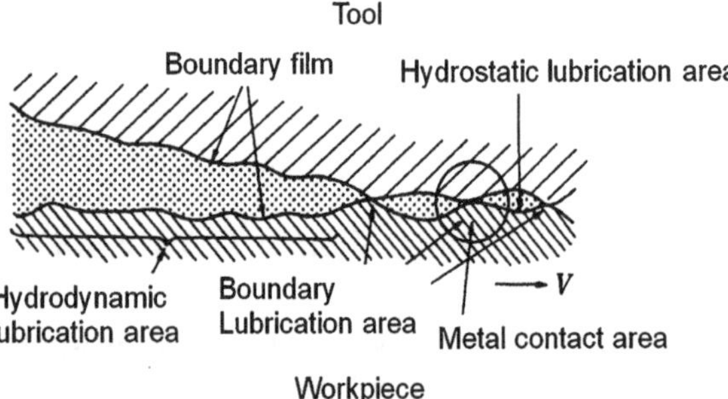

Fig. 5.9 Schematic lubrication model at interface between tool and workpiece in metalforming

5.2.3 *Micro-Contact Under Micro-Plasto-Hydrodynamic Lubrication*

Generally, a schematic lubrication model at the interface between tool and workpiece in metal forming is shown in Fig. 5.9. In this model, it is explained that the lubrication between tool and workpiece consists of a boundary region in which the workpiece surface under the tool is dressed and a hydrostatic lubrication region in which the lubricant is filled in the small pockets on the workpiece surface.

In this hydrostatic-boundary lubrication regime, a part of the normal contact load is supported by the hydrostatic pressure q generated in the surface pockets of the workpiece, and the remainder is supported by the asperities, as shown in Fig. 5.10. Then, in this lubrication condition the tool having a smooth surface moves in the tangential direction at a speed of V, so that we can observe that the trapped lubricant within the surface pockets permeates into the real contact area.

Therefore, Azushima et al. have tried to observe directly the permeation of the lubricant from the pocket to the real contact area [6]. They used a newly developed drawing apparatus. Figure 5.11 shows the schematic representation of the apparatus for direct observation of the lubricant permeation.

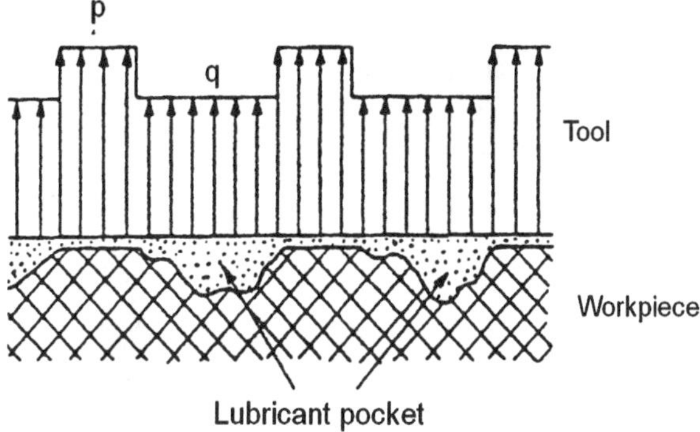

Fig. 5.10 Model of hydrostatic-boundary lubrication

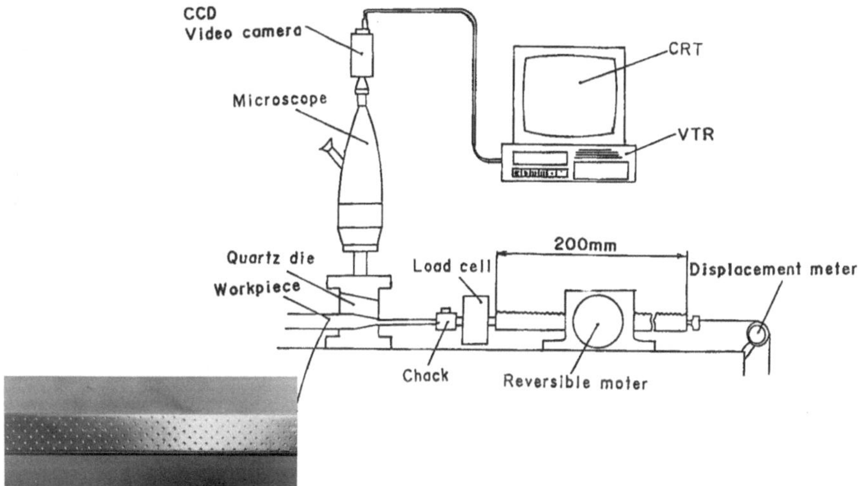

Fig. 5.11 Schematic representation of apparatus for direct observation

The experimental apparatus consists of a transparent die made of quartz, a microscope with a CCD video camera and a video system. The specimen is clamped with the edge of linear head is drawn by a reversible motor of 20 W and the drawing length is 20 cm. The drawing speed can change within a range from 0.1 to 10 mm/s. The die angle is 3°. The workpiece used is A1100 Aluminum, 2 mm thick, 10 mm wide and 300 mm long. The specimen sheets are provided with uniformly distributed pyramidal indentation as shown in this figure. The distance between the pyramidal indentations is 2 mm and the angle is 160°. The geometries of indentation are 0.63 mm length and 0.056 mm depth. Drawing experiments are carried out at two speeds of 0.2 mm/s and 0.8 mm/s changing the reduction in the range from 3 % to 13 %. Three Paraffinic oils with different viscosities are chosen as lubricant.

Firstly, the experimental results of the geometrical change of the pyramidal indentation is shown [6]. Figure 5.12a illustrates the sequence photographs of the

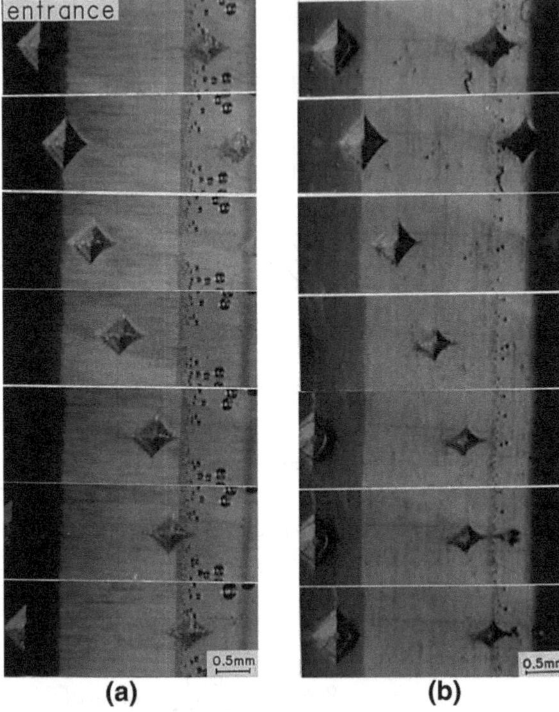

Fig. 5.12 Sequence photographs of surface of Aluminum sheet having pyramidal pockets being drawn at a drawing speed 0.2 mm/s at a reduction of 7.1 % (**a**) and 9.0 % (**b**) with oil having 100 cSt

working area taken during drawing at a drawing speed of 0.2 mm/s at a reduction of 9.0 % with a lubricant having 100 cSt viscosity. When the lubricant is completely trapped in the surface pocket of the pyramidal indentation in the working area, the geometry observed does not change hardly compared with one before drawing. However, when the air in the pocket is introduced before the working area as shown in Fig. 5.12b, the geometry of indentation decreases largely with the deformation of workpiece.

From these results, it will be seen that the hydrostatic pressure generated in the lubricant trapped completely in the pocket so that the indentation geometry in the working area does not change largely.

Secondly, the experimental results on the behavior of the lubricant trapped in the pocket are shown. From the detail observation of Figs. 5.13a and 5.13b, we can understand that the lubricant trapped in the pocket is permeated to the real contact area. We can show the photographs of the working area taken during drawing, which clearly show the lubricant squeezed out from the oil pocket. Figure 5.13a illustrates the sequence photographs of the working area taken during drawing with mixture lubricant added 4 % red paint to Paraffinic oil having a viscosity 100 cSt at a speed of 0.8 mm/s at a reduction of 9.7 %.We can observe that the lubricant in the pocket permeates forward and backward at a same time into the real contact area. The permeation area on the real contact area extends with reduction. As you can see, the amount of the forward permeated lubricant from the

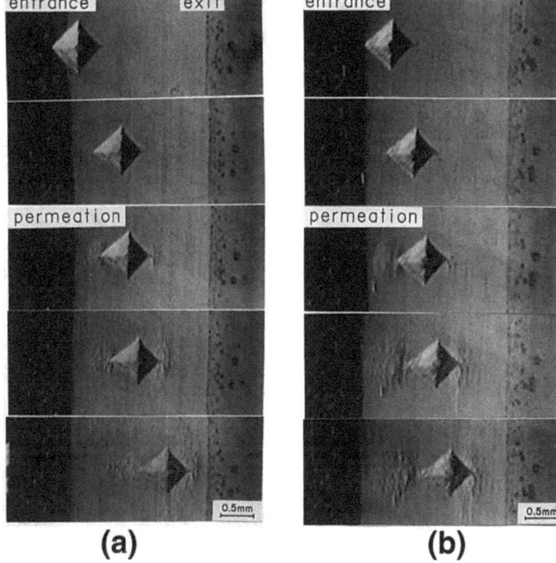

Fig. 5.13 Sequence photographs of permeation behavior of lubricant trapped in pyramidal pockets during sheet drawing at a drawing speed 0.8 mm/s at a reduction of 9.7 % with oil having 100 cSt (**a**) and at a reduction of 9.2 % with oil having 1000 cSt (**b**)

pocket is larger than one of the backward permeated lubricant. Figure 5.13b illustrates the sequence photographs of the working area taken during drawing with the mixture lubricant added 8 % red paint to Paraffinic oil having a viscosity of 1000 cSt at a speed of 0.8 mm/s at a reduction of 9.2 %. In this figure, for the lubricant having higher viscosity, the lubricant in the pocket seems to be squeezed much more backward into the real contact area. From these direct observations, it is seen that the permeation behavior of lubricant in the pocket depends considerably on the viscosity of lubricant.

Next, the behavior of lubricant permeation will be considered qualitatively. Figure 5.14 shows the schematic representation of two dimensional model of the lubricant permeation.

It is anticipated from the assessment of the lubricant pressure p_s in the surface pocket and the observation of the marked volume shrinkage of the pocket of pyramidal indentation shown in Fig. 5.12, that p_s reaches close to the pressure $p_{r(x)}$ on the real contact area when the workpiece undergoes bulk deformation. Moreover, it is well expected that pd rises at the back margin on he pocket by the hydrodynamic effect depending on the product of viscosity of lubricant and drawing speed ηV.

Figure 5.14a shows the general relation among $p_{r(x)}$, ps and pd when the pocket exists area. The hydrodynamic pressure p_d can be derived by the Reynolds equation. $p_{r(x)}$ generally decreases toward the tool exit in the drawing process, but ps increases with the movement of the pocket from the tool entrance to the tool exit. It is well expected that $p_{s2} + p_{d2}$ is like to for each p_{r2} first, see Fig. 5.14b when ηV is high, thus resulting in backward permeation of the trapped lubricant. This model is in good agreement with the experimental results as shown in Fig. 5.13b.

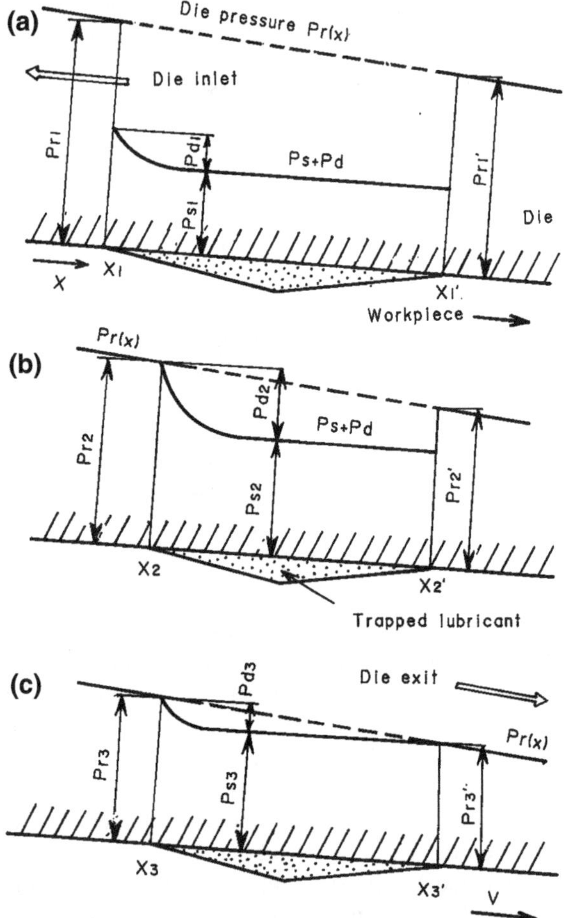

Fig. 5.14 **a** Models to account for lubricant for lubricant permeation from surface pocket into real contact surface area **b** Backward permeation **c** Backward and forward permeation

On the other hand, it is expected that p_{s3} is to reach p_{r3} at the front margin of the pocket, when/before $p_{s3} + p_{d3}$ is to reach p_{r3} at the back margin when the viscosity or the speed is low, see Fig. 5.14c, thus resulting in forward permeation or backward permeation.

In this lubrication mechanism, the tubes with a mirror surface of aluminium alloys is produced by ironing process for the production of photo-sensitive drums for copying machines. An example of industrial success obtained by the use of the micro-hydrodynamic lubrication is shown in Fig. 5.15. The surface roughness is less than 0.2 μm Rmax. In this manufacturing process, the cold backward extruded performs are etched by an alkaline solution to attain a roughness of 1 to 3 μm Rmax. They are then ironed through a mirror finished tungsten-carbide die with a certain kind of oil mixture.

Fig. 5.15 Cold extruded and ironed photo-sensitive drums of aluminum alloy

5.2.4 Micro-Contact Under Mixed Lubrication

Generally, the mixed lubrication regime consists of the hydrodynamic lubrication and the boundary lubrication or the hydrostatic lubrication and the boundary lubrication. Figure 5.16 shows the model of the mixed lubrication of the hydrodynamic lubrication and the boundary lubrication. Figure 5.17 shows the model of the mixed lubrication of the hydrostatic lubrication and the boundary lubrication.

In Fig. 5.16, the real contact area ratio α is given by the Eq. (5.6) and moreover, the shear stress τ is given by the Eq. (5.7).

$$\alpha = \int_h^\infty \frac{1}{\sigma\sqrt{2\pi}} \exp\left(-\frac{z}{2\sigma^2}\right) dz \tag{5.6}$$

$$\tau = (1 - \alpha)\eta \frac{\partial u}{\partial y} + \alpha \mu_b p \tag{5.7}$$

In Fig. 5.16, the shear stress is given by the Eq. (5.8).

$$\tau = (1 - \alpha)\mu_b p \tag{5.8}$$

In this lubrication mechanism, the sheet with a mirror surface of stainless steel is produced by multi-pass cold rolling. In this rolling, the surface texture of the rolled sheet is controlled using rolls with regular surface profiles and in the final pass the sheet with a mirror surface of stainless steel as rolled [7].

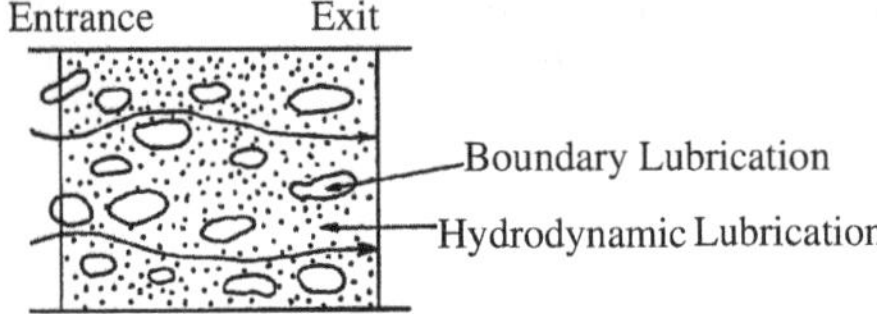

Fig. 5.16 Model of hydrodynamic-boundary lubrication

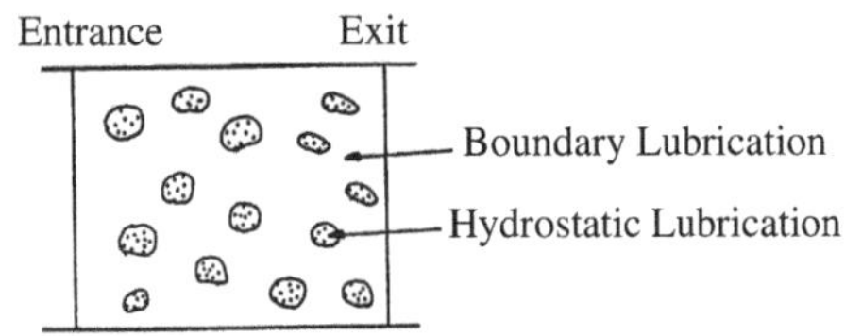

Fig. 5.17 Model of hydrostatic-boundary lubrication

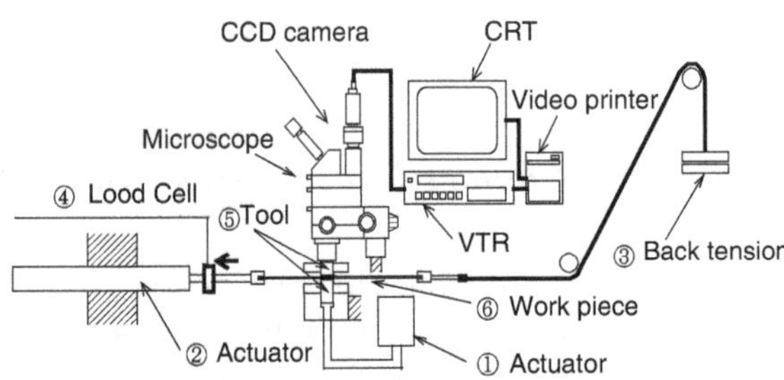

Fig. 5.18 Schematic representation of apparatus for in situ observation

5.3 Micro-Contact at the Interface Between Tool and Workpiece in Sheet Metal Forming

5.3.1 Direct Observation of Micro-Contact

From the dire observation of the micro-contact at the interface between tool with smooth surface and workpiece with random surface, we tried to investigate into the lubrication regime and the coefficient of friction [8].

Figure 5.18 shows the schematic representation of an apparatus for in situ observation of the interface between tool and workpiece. The experimental apparatus consists of a pair of tool halves having flat surfaces, one made of high speed steel and the other transparent quartz, a microscope with a CCD camera and a video system. The model workpiece material is commercially available pure annealed A1100 aluminum sheets, 1 mm thick, 10 mm wide and 500 mm long. The aluminum sheets are rolled at a constant reduction of 5 % with a light mineral oil, using shot blast dull rolls in order to provide an intentionally high surface roughness Rmax 20 μm on the specimen surface. The proof stress is 64 MPa. The flat tool sheet drawing experiments are carried out at a constant speed of 8.4 mm/min at 6 different average contact pressures within the range from 19 to 56 MPa. Paraffinic base oil having a viscosity of 1480 cSt with 5 % oleic acid is used as a lubricant.

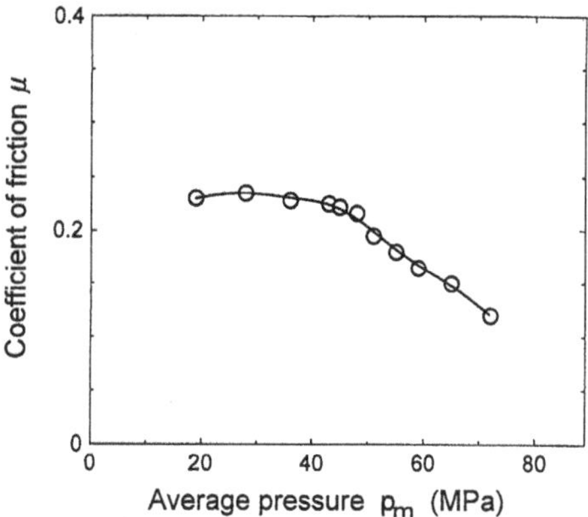

Fig. 5.19 Relationship between nominal coefficient of friction and average contact pressure

In each experiment, the normal force and the drawing force are measured to determine the nominal coefficient of friction. At the same time, video photographs of the micro-contact with the CCD camera.

Figure 5.19 illustrates the relationship between nominal coefficient of friction and average contact pressure. The nominal coefficient of friction remains constant at 0.23 up to a average contact pressure of 41 MPa and decreases from 0.23 to 0.12 below 41 MPa.

In Fig. 5.20, the interface appearances at six different average contact pressures from 19 to 72 MPa are compared. At the low average contact pressure of 19 and 36 MPa, the real contact zones are isolated. The average area of the individual zones and the sum of the real contact area becomes larger as the pressure increases. The video photographs clearly show the free lateral flow of the lubricant under such low contact pressures, As the normal contact pressure increases to 41 MPa, the real contact zones grow and some of them start to connect so as to form closed lubricant pools in which no lateral flow is observable. Beyond 41 MPa, the number of isolated lubricant pockets increases gradually with increasing average contact pressure. As the average contact pressure increases to 59 MPa, the bright area of real contact suddenly becomes dull, suggesting that a kind of micro-roughening takes place. The contact area ratio at this stage is about 70 %.

5.3.2 Lubrication Mechanism at Interface Between Tool and Workpiece

The constant nominal coefficient of friction in the low pressure from 19 to 41 MPa, where the lubricant flows freely out through the interface gap and accordingly it is understood that the hydrostatic pressure within the lubricant is not generated the lubrication mechanism can be accounted for the boundary lubrication mechanism as shown in Fig. 5.21.

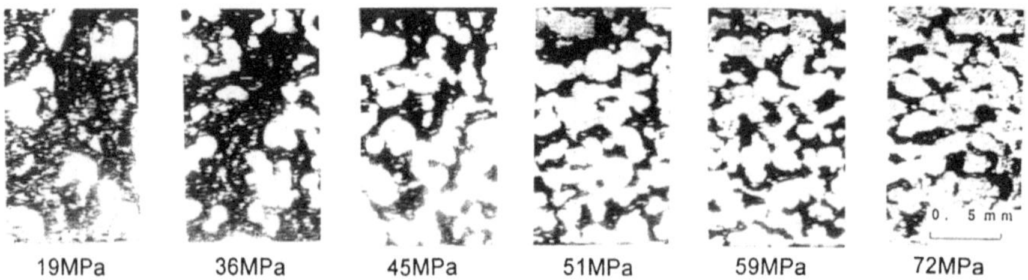

Fig. 5.20 Video images of test aluminium workpiece for six different average contact pressure

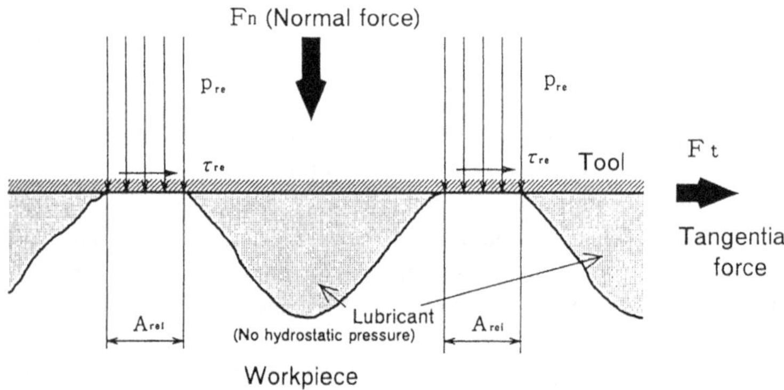

Fig. 5.21 Schematic representation of contact model between tool and workpiece in boundary lubrication

Beyond he average contact pressure of 41 MPa, the nominal coefficient of friction decreases with increasing average contact pressure pm. In the pressure ranges from 41 MPa to 51 MPa, it is confirmed from the CRT image of the interface that some of the real contact zones start to connect so s to form closed lubricant pools in which no lubricant flow is observable. Butlur [9] first suggested that if liquid lubricant is introduced at the interface, the lubricant is trapped within the pocket on the workpiece surface. Also, Kudo [10] and Wanheim et al. [11] theoretically assessed the hydrostatic pressure q generated within the lubricant trapped in the surface pocket. The schematic representation of this hydrostatic-boundary lubrication is shown in Fig. 5.22,

Lastly, above the average contact pressure of 59 MPa, it is observed in the CRT image that the bright zone of real contact suddenly becomes dull, suggesting that a kind of micro-roughening takes place and that the lubricant trapped in the pocket permeates into the real contact area. In order to describe this phenomena, Mizuno et al. [12] proposed the concept of micro-plasto-hydrodynamic lubrication. Later, Azushima et al. [6] confirmed this hypothesis by observing the surface during sheet drawing through a transparent tool. The mechanism also supports the hypothesis that q approaches pre. The schematic representation of the lubrication mechanism is shown in Fig. 5.23.

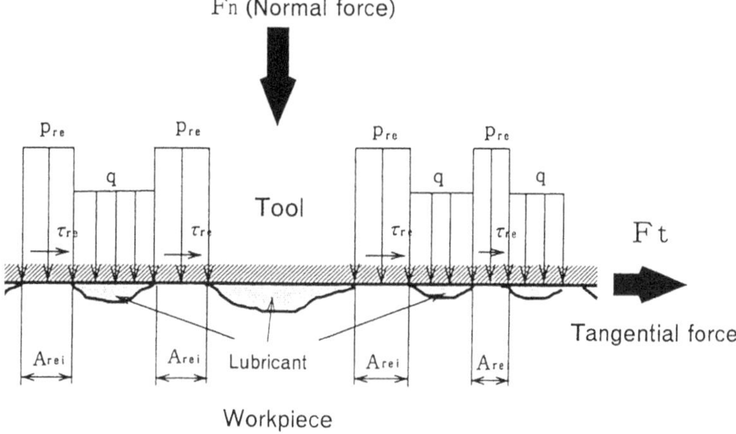

Fig. 5.22 Schematic representation of contact model between tool and workpiece in hydrostatic-boundary lubrication

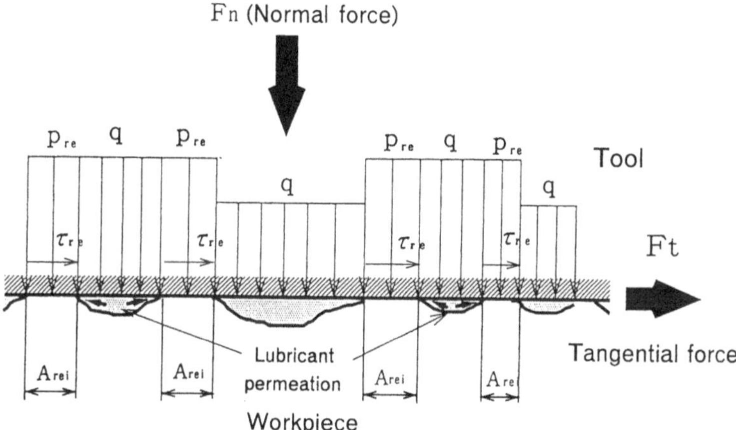

Fig. 5.23 Schematic representation of contact model between tool and workpiece in boundary-hydrostatic-micro-PHL lubrication

5.4 Micro-Contact at Interface Between Roll and Sheet in Cold Sheet Rolling

5.4.1 Inlet Oil Film Thickness

The inlet oil film thickness in a steady process as rolling becomes a useful parameter in order to estimate the lubrication behavior. Figure 5.24 shows the schematic representation of the inlet zone in the roll bite. The roll and the sheet are

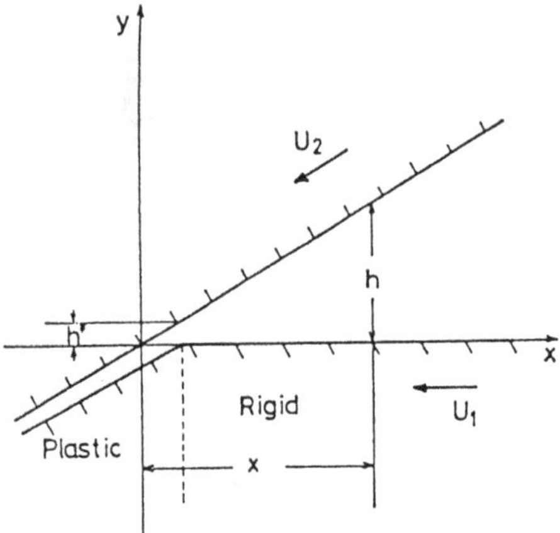

Fig. 5.24 Schematic representation in inlet zone between roll and sheet

rigid in the inlet zone. When the pressure in the lubricant builds up to the yield stress, the sheet deforms plastically.

Reynolds equation is given by

$$\frac{dp}{dx} = 6\eta(U_1 + U_2)\left(\frac{h - h^*}{h^3}\right) \tag{5.9}$$

where η is the viscosity and h* is the inlet oil film thickness.The lubricant viscosity is assumed to obey the equation of the form

$$\eta = \eta_0 \ \exp\{\alpha p - \beta(T - T_0)\} \tag{5.10}$$

where η_0 is the viscosity of lubricant in the ambient pressure and temperature, α and β are the pressure and temperature coefficient of viscosity respectively and T_0 is the ambient temperature.

The inlet oil film thickness at the interface between roll and sheet considering the dependence of the lubricant viscosity on the pressure and the temperature is calculated. Then, the assumptions employed in the analysis are as follows. The heat convected by the lubricant is neglected, the surface of roll is isothermal and is equal to the ambient temperature T_0 and the viscosity of lubricant is constant across the film and is a function of the temperature and the pressure.The energy equation is given by

$$K\frac{\partial^2 T}{\partial y^2} + \eta\left(\frac{\partial u}{\partial y}\right)^2 = 0 \tag{5.11}$$

The inlet oil film thickness is numerically calculated using Eqs. (5.9)–(5.11). The calculated results of the inlet oil film thickness are shown in Fig. 5.25 [13]. The inlet oil film thickness considering the effect of the lubricant viscosity is smaller than the isothermal that. In particular, the difference between them becomes bigger in a higher rolling speed. If the rolling speed increases, the inlet oil

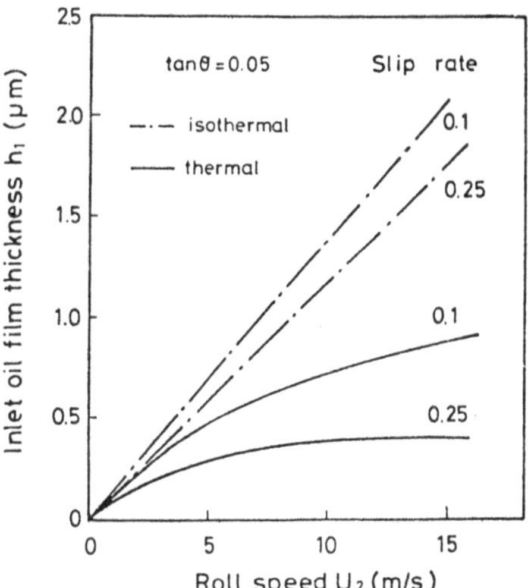

Fig. 5.25 Relationship between calacutared inlet oil film thickness and roll speed

film thickness becomes thinner and it can be considered that the metal pick up on the roll surface will occur.

5.4.2 Micro-Contact at Interface Between Roll and Sheet

The micro-contact at the interface between roll and sheet is affected by many tribological factors such as the operation factors (rolling speed V_r, inlet speed of sheet V_1, inlet oil film thickness h*, reduction in thickness r, roll radius R and flow stress of sheet Y), the lubricant factors (viscosity η_0, pressure coefficient α and temperature coefficient β) and the surface roughness of roll and sheet (R_{a0}^{roll},R_{a0}^{sheet}). However, it is well-known that the micro-contact strongly depends on the inlet oil film thickness and the surface texture of strip after rolling [5].The inlet oil film thickness can be calculated using Eqs. (5.9)–(5.11). On the other hand, the surface texture is estimated from the surface brightness of sheet measured after rolling.

Figure 5.26 shows the surface appearances of sheet after rolling. The rolling experiments are carried out changing the rolling speeds from 1.2 to 121 m/min and the viscosities of lubricant from 8 to 97.4 cSt (at 20 °C) at a constant reduction in thickness using steel sheet and SUJ2 roll of 76 mm in diameter with a mirror surface [14]. Figure 5.27 shows the relationship between surface brightness and rolling speed [14]. It is understood that the surface brightness Gs depends on the many tribological factors. Consequently, the surface brightness Gs of the rolled sheet can be expressed as a next equation.

$$Gs = Gs\left(V_r, V_1, h^*, r, R, k, \eta_0, \alpha, \beta, R_{ao}^{roll}, R_{a0}^{sheet} \right) \quad (5.12)$$

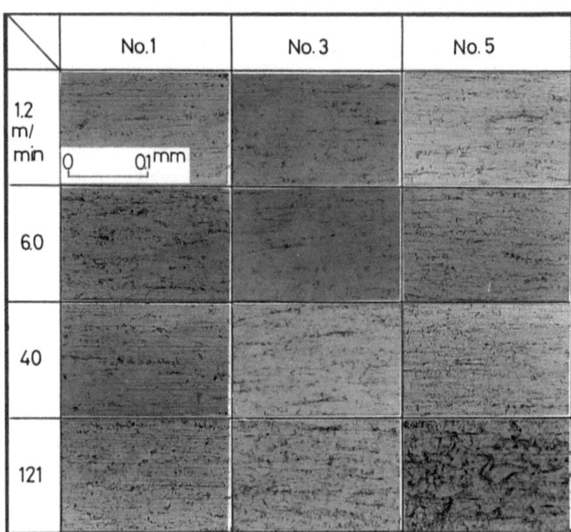

Fig. 5.26 Surface photographs of rolled sheet at rolling speeds of 1.2, 6.0, 40 and 121 m/min at a rolling reduction of 20 % using lubricants of No.1, No.3 and No.5

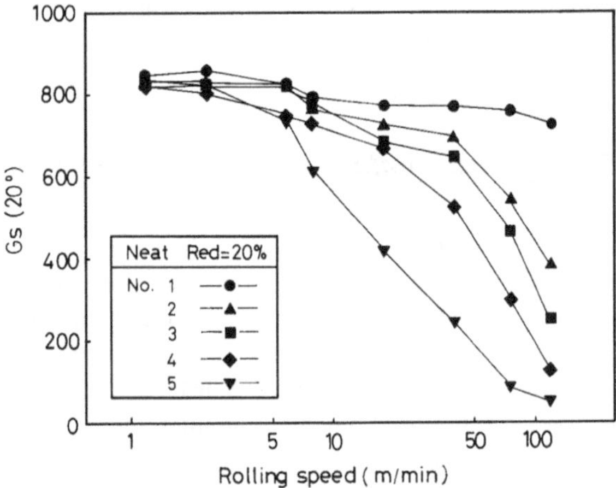

Fig. 5.27 Relationship between surface brightness and rolling speed at a rolling reduction of 20 % using lubricants of No.1 to No.5 when the surface roughnesses of roll and sheet are 0.01 and 0.07 μmRa

Then, Fig. 5.28 shows the relationship between surface roughness Gs and inlet oil film thickness. The inlet oil film thicknesses are calculated using the Eqs. (5.9)–(5.11) [7]. The surface brightness Gs uniquely depends on the inlet oil film thickness h* as shown in Fig. 5.27, when the surface roughnesses of roll and sheet are smooth.

In the region of the inlet oil film thickness up to 10^{-2} μm, the surface brightness is constant and it is assumed that the lubrication mechanism becomes boundary. In the region over about 0.4 μm, the surface brightness remains lower and the lubrication mechanism becomes hydrodynamic. In the middle region, the surface brightness decreases linearly with increasing inlet oil film thickness and the

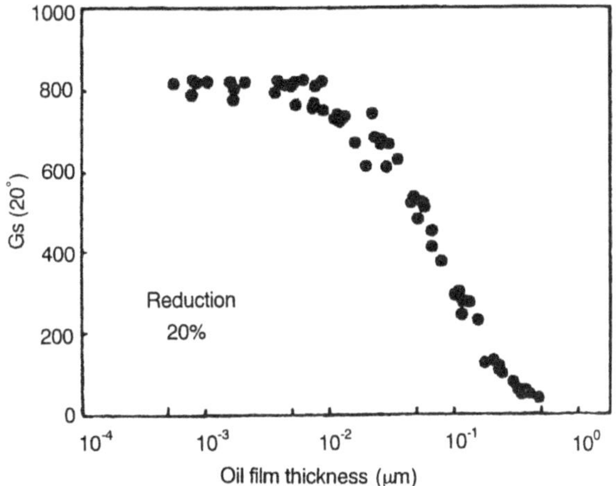

Fig. 5.28 Relationship between surface brightness shown in Fig. 5.27 and inlet oil film thickness

lubrication mechanism becomes mixed. From these results obtained, the surface brightness Gs is expressed as a next equation.

$$Gs = Gs(h^{*}) \tag{5.13}$$

However, since the surface brightness is affected by the surface roughness of roll and sheet before rolling, it is expressed as three functions of the inlet oil film thickness and the surface roughnesses of roll and sheet as a following equation.

$$Gs = Gs\left(h^{*}, R_{a0}^{roll}, R_{a0}^{sheet}\right) \tag{5.14}$$

5.4.3 Estimation System for Micro-Contact of Rolled Sheet in Stainless Steel Cold Rolling Process

If the Eq. (5.14) becomes complete, a system for estimating the surface brightness after rolling will be constructed using the relation. Consequently, the micro-contact at the interface between roll and sheet can be estimated by three functions of the inlet oil film thickness and the surface roughness of roll and sheet. However, since it is difficult to obtain quantitatively the complete relation from the experimental data base, another system must be constructed in order to estimate the surface brightness in the actual mill operation.

The new system for cold rolling of stainless steel is proposed [7]. The flow chart of the system is shown in Fig. 5.29. Experiments for the estimation are carried out as follows;

(I) Preparation of as-hot-rolled sheet,
(II) Input of rolling conditions at each pass in actual mill,
(III) Calculation of the inlet oil film thickness at each pass,

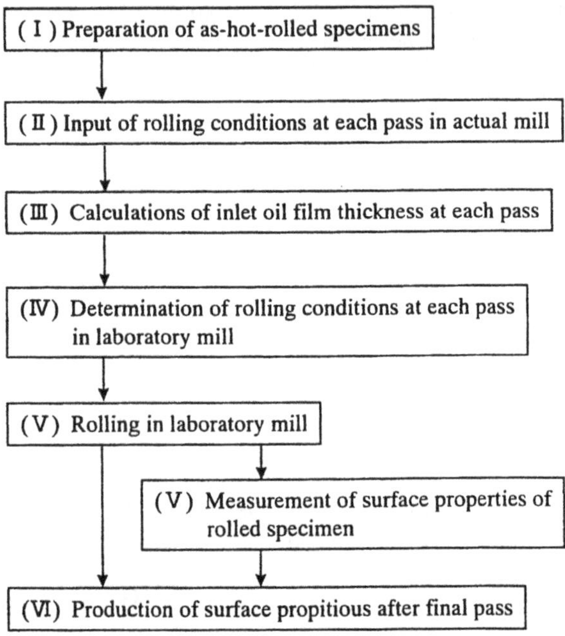

Fig. 5.29 Flow chart of estimation system for surface brightness of rolled sheet

Table 5.1 Rolling conditions in A company

Pass		1	2	3	4	5	6	7	8	9
Rolling condition										
Thickness (mm)	4.0	3.5	2.85	2.27	1.78	1.40	1.10	0.87	0.70	0.60
Reduction (%)		12.5	18.6	20.4	21.6	21.3	21.4	20.9	20.7	14.5
Front tension (Kgf/mm^2)		10	12	14	18	19	23	27	31	31
Beck tension(Kgf/mm^2)		–	6	7	9	14	16	18	24	27
Rolling speed (m/min)		170	250	250	250	250	250	210	200	150
Oil temp (°C)		40	40	40	40	40	40	40	40	40

(IV) Determination of rolling conditions at each pass in laboratory mill,
(V) Simulation rolling in laboratory mill and measurement of surface properties of rolled specimen,
(VI) Output of surface brightness after final pass.

In order to evaluate the new system for estimating the surface brightness of rolled stainless steel before rolling, a stainless steel strip is rolled in A steel making company. Table 5.1 shows the rolling conditions in A company. In A company, rolling is carried out at 9 passes with the thickness reduction of from 4.0 mm to 0.60 mm. The rolling speeds are 170 m/min at 1 pass, 250 m/min at 2 to 7 passes, 200 m/min at 8 pass and 150 m/min at 9 pass.

Table 5.2 Rolling conditions of simulation tests in A company

	1Pass	2Pass	3Pass	4Pass	5Pass	6Pass	7Pass	8Pass	9Pass
Roll Material	SKD								SKH
Diameter (mm)	76.0								
Surface roughness of roll (Raμm)	0.317							0.166	0.030
Inlet angle (°)	6.58	7.51	7.10	6.52	5.74	5.10	4.46	3.84	2.94
ROiling speed (m/s)	0.085	0.127						0.098	0.075
Viscosity of lubricant (Pa s)	0.27								
Intel oil film thickness (μm)	0.0210	0.0270	0.0285	0.0309	0.0352	0.0397	0.0454	0.0409	0.0414

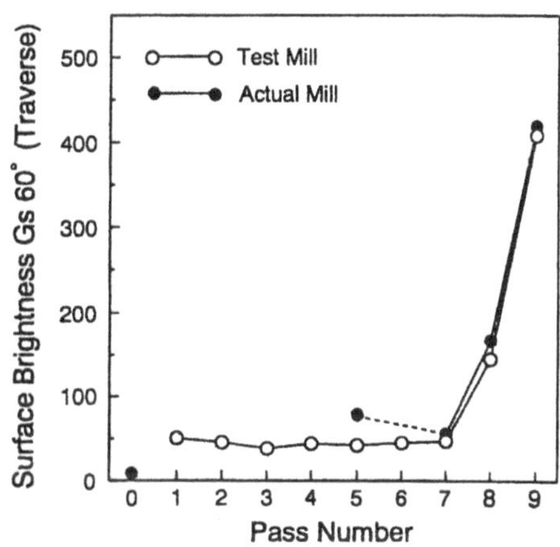

Fig. 5.30 Comparison with surface brightness obtained by test mill and one by actual mill

The rolling simulation tests are carried out using laboratory mill with a roll of a radius 76 mm. Table 5.2 shows the estimation conditions of the laboratory mill. The roll materials used are SKD11 and SKH(JIS). The sheet is cut off from the as-hot-rolled strip is used. The sheets have a width of 20 mm and a length of 250 mm and the thicknesses in each pass are shown in Table 5.2. The sheet surface is degreased with Hexane before rolling. After rolling using the laboratory mill, the surface brightness is measured using a brightness meter.

Figure 5.30 shows the comparison with the surface brightnesses measured by the actual mill and by the simulation rolling test. A good agreement can be also obtained between them. The surface appearances of rolled sheet at 5, 7, 8 and 9 passes are shown in Fig. 5.31.

From these results, the micro-contact at the interface between roll and sheet in the each pass of the cold actual mill of stainless steel can be estimated.

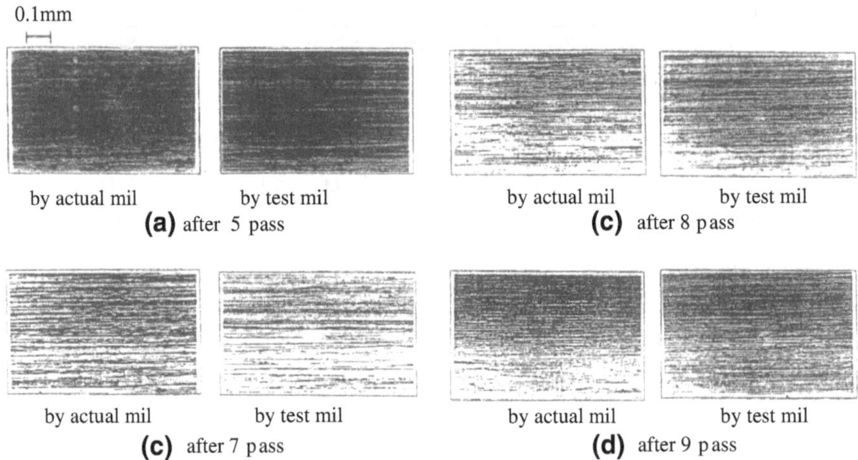

Fig. 5.31 Surface appearances of rolled strips and sheet at 5, 7, 8 and 9 passes

5.5 Conclusions

The micro-contacts at the interface between tool and workiece under each lubrication are reviewed. In the micro-contact under dry and thin film lubrication (boundary lubrication), the sheet with a mirror surface is produced by skin pass rolling. However, in this rolling process the productivity with higher rolling speed can not raised due to occurrence of friction pick up. In the micro-contact under micro-plasto hydrodynamic lubrication, the tubes with mirror surface of aluminum alloys is produced by ironing process for the production of photo-sensitive drums for copy machines. In the micro contact under mixed lubrication, the sheet with a mirror surface of stainless steel is produced by multi-pass cold rolling.

The micro-contact behavior at the interface between tool and workpiece observed by means of a newly developed drawing apparatus which consisted of a transparent tool made of quartz is explained. In the lower average pressure, the coefficient of friction is constant and the regime is characterized by isolated contact area according to the boundary lubrication model. In the medium range, the coefficient of f friction decreases with increasing pressure and the regime is characterized by closed lubricant pools in which the hydrostatic is generated according to the hydrostatic-boundary lubrication model. In the higher pressure, the coefficient of friction decreases also increasing normal pressure and the regime is characterized by oil permeation into the real contact area according to the boundary, hydrostatic and micro-plasto-hydrodynamic lubrication model.

The micro-contact at the interface between roll and sheet in stainless steel cold rolling process is explained using the system for estimating the surface brightness from the relationship between the inlet oil film thickness and the surface brightness of rolled sheet. In this rolling process, the surface texture of the rolled sheet is controlled using rolls with regular surface profiles and in the final pass the sheet with a mirror surface of stainless steel is rolled. The micro-contact at the interface between roll and sheet after each pass in actual mill can be estimated using this system.

References

1. Kudo H, Azushima A (1987) Interaction of surface microstructure and lubricant in metal forming tribology. Proc Second ICTP 373–385
2. McFarlane JS, Tabor D (1950) Relation between friction and adhesion. Proc Roy Soc London A202:244–253
3. Bay N, Wanheim T (1976) Real area of contact and friction stress at high pressure sliding contact. Wear 38:2101–2209
4. Makinouchi A et al (1986) Flattening mechanism the pressure of bulk plastic deformation of workpiece. In:Proceeding 1986 Japanese Spring conference Technology of Plasticity, Tokyo 127–130
5. Azushima A et al (1978) An analysis and measurement of oil film thickness in cold sheet rolling. J Japan Soc Technol Plasticity 19:958–965
6. Azushima A et al (1990) Direct observation of lubricant behavior under the micro-PHL at the interface between workpiece and die. Ann CIRP 39:551–556
7. Azushima A (1998) Estabilishment of estimation system for surface brightness of rolled sheet in stainless steel cold rolling process. In: Proceeding 7th international Confeference on Steel Rolling 473–478
8. Azushima A (1995) Direct observation of contact behavior to interpret the pressure dependence of the coefficient of friction in sheet metal forming. Ann CIRP 44:209–212
9. Butler LH (1960) Surface conformation of metals under high normal contact pressure. Metallurgia 58:167–174
10. Kudo K (1965) A note on the role of microscopically trapped lubricant at the tool-work interface. Int J Mech Sci 7:383–388
11. Wanheim T, Bay N (1978) A model for friction in metal forming processes. Ann CIRP 27:189–194
12. Mizuno T, Okamoto M (1982) Effects of lubricant vissosity at pressure and sliding velosity of lubricating conditions in the conpression-friction test on sheet metals, J Lubr Technol ASME 104:53–59
13. Azushima A (1987) Lubrication in steel strip rolling in Japan. Tribol Int 20:316–321
14. Azushima A, Noro K, Inagaki Y, Degawa H (1990) Proposal of control system of surface brightness of rolled sheet in cold rolling, Tetsu-to-Hagane, 76:576–583

Chapter 6
Coating and Applications

Cássio A. Suski and C. A. S. de Oliveira

Abstract This chapter presents the development of new coatings in order to minimize the effects of wear on the tools of manufacturing processes. The coating materials with hard layers results in increased wear resistance and decreasing the coefficient of friction, allowing a better tribological performance of tools and a reduction in machine downtime due to premature failure. The Plasma nitriding, the Chemical Vapor Deposition (CVD), the Physical Vapor Deposition (PVD), the Physical–Chemical Vapor Deposition (PCVD), some thermochemical treatments, multilayer coatings and Diamond-Like Carbon (DLC) were discussed. The physical vapor deposition can be obtained through three technical features: deposition by vacuum evaporation, cathode sputtering and ion plating. Some coatings properties, such as high hardness, low friction coefficient and high chemical and thermal stability, are primarily responsible for the diversity of applications of these coatings. In this chapter are also presented and compared the properties of various surface coatings applied in the industry.

This chapter presents the development of new coatings in order to minimize the effects of wear on the tools of manufacturing processes. The coating materials with hard layers results in increased wear resistance and decreasing the coefficient of friction, allowing a better tribological performance of tools and a reduction in machine downtime due to premature failure. The Plasma nitriding, the Chemical Vapor Deposition (CVD), the Physical Vapor Deposition (PVD), the Physical–Chemical

C. A. Suski (✉)
Instituto Federal de Santa Catarina, 3122 St n. 340 apt. 901,
Balneário Camboriú, SC 88330-290, Brazil
e-mail: cassio_suski@hotmail.com

C. A. S. de Oliveira
Mechanical Engineering Department, Universidade Federal
cde Santa Catarina, Florianopolis, SC, Brazil
e-mail: carlosa@emc.ufsc.br

J. P. Davim (ed.), *Tribology in Manufacturing Technology*, Materials Forming, Machining and Tribology, DOI: 10.1007/978-3-642-31683-8_6,

Vapor Deposition (PCVD), some thermochemical treatments, multilayer coatings and Diamond-Like Carbon (DLC) were discussed. The PVD can be obtained through three technical features: deposition by vacuum evaporation, cathode sputtering and ion plating. Some coatings properties, such as high hardness, low friction coefficient and high chemical and thermal stability, are primarily responsible for the diversity of applications of these coatings. In this chapter are also presented and compared the properties of various surface coatings applied in the industry.

6.1 Introduction

In previous decades the need to reduce the costs of production processes imposed by the industrial development, coupled to the increasing competitiveness based on quality, led research centers to study new materials and coatings in order to minimize the effect of wear on tools. Coating materials with hard layers increases their resistance to wear and decreases their coefficient of friction, allowing for a better tribological performance of tools and a reduction in machine downtime due to premature failure.

Tools with superficial coatings resistant to wear are increasingly used due to the important improvement of their performance based on the reduction of friction and plastering. The CVD on industrial scale began in the 1960s when it was designed for the deposition of wear resistant surface coatings such as TiCN (titanium carbonitride) and TiN (titanium nitride) on hard metals. In the 1970s, a process of PVD was developed simultaneously by both Balzers in Europe and ULVAC and Sumitomo in Japan. Due to the use of low temperatures, this technique allowed coating high speed steel tools without losing the hardness of the matrix and with no resultant need for a new quenching heat treatment. Besides the techniques already mentioned, other surface treatment/coating techniques such as electrodeposition, nitriding, boriding, black oxide coating, PCVD and DLC are currently used to increase the durability of mechanical forming tools [1].

In mechanical forming operations, coatings provide better finish for the final product, as well as reduction of efforts and decrease of tools wear.

During the coating of metal forming tools, in which the main purpose is generally to form a superficial film to promote the increase of wear resistance, the vapor deposition processes are the most commonly used. Vapor deposition processes can be divided basically into three major groups: CVD, PVD and PCVD [1]. CVD uses a volatile component of the coating which is thermally decomposed (pyrolysis) or reacts chemically with other gases or vapors to form a coating on the heated surface of the substrate [2].

PVD is used to deposit coatings by means of high vacuum condensation of vapors on the surface of the substrate. The deposition occurs atomically, i.e. atom by atom.

PCVD is a hybrid process that uses a glow discharge to activate the CVD process.

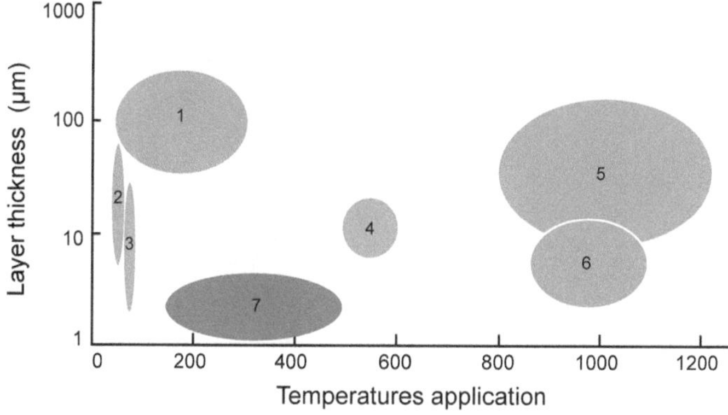

Fig. 6.1 Layer thickness and application temperatures of some surface modification techniques—Plasma spray (1); Electrolytic deposition (2); Phosphating (3); Nitriding (4); Boriding (5); Chemical Vapor Deposition—CVD (6) and Physical Vapor Deposition—PVD (7)

Table 6.1 Major thermochemical treatments and their applications

Treatment	Definition	Application
Boriding	Thermochemical treatment which produces superficial enrichment with boron	Parts requiring high abrasion resistance
Carbonitriding	Thermochemical treatment which produces superficial enrichment both with carbon and nitrogen	Parts requiring high superficial hardness, high resistance to contact fatigue, and which are subjected to moderate superficial charges
Carburizing	Thermochemical treatment which produces superficial enrichment with carbon	Parts requiring high superficial hardness, high resistance to contact fatigue, and which are subjected to high superficial charges.
Nitriding	Thermochemical treatment which produces superficial enrichment with nitrogen	Parts requiring high resistance to adhesive friction, high resistance to contact fatigue, and which are subjected to low superficial charges

Figure 6.1 presents a comparison with the layer thicknesses and application temperatures of some surface modification techniques applied to metal forming tools.

Besides the processes of coating deposition on the working surface of the tools, thermochemical treatments are also widely used to increase the durability of punches and dies for metal forming.

Thermochemical treatments can be defined as a set of operations performed in the solid state, which consist of modifications in the chemical composition of the parts surface at appropriate temperature and environment [3]. Table 6.1 presents the definitions and general applications of the most widely used treatments.

General aspects of plasma nitriding, CVD and PVD are presented below.

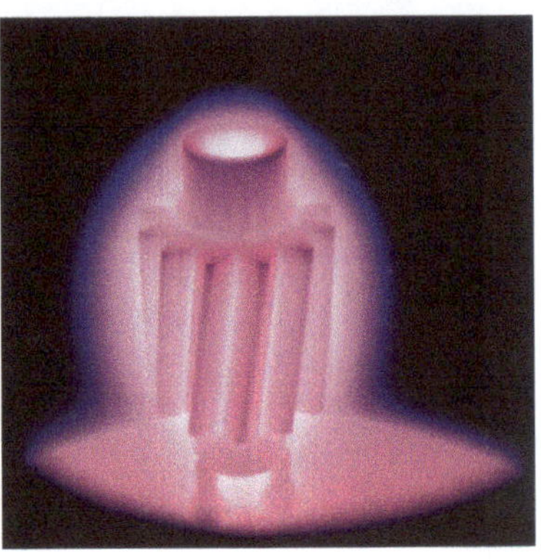

Fig. 6.2 Component under plasma nitriding

6.2 Plasma Nitriding

The nitriding treatment is performed at low temperatures, where the maximum solubility of nitrogen in the matrix is 0.1 % of mass. When the nitrogen content exceeds this value, there is a formation of nitrides. The first nitride to be formed is γ-Fe_4N, which is tetragonal and stable up to about 6.1 % of mass of nitrogen. Above this level, there is a formation of the nitride ε-$Fe_{2\text{-}3}N$, hexagonal. In general, the nitrided surface is viewed under optical microscope as presenting a layer of compounds in the outer surface, called white layer, followed by the diffusion zone, which consists of layers around 300 μm.

The nitriding processes traditionally used in industry are gas nitriding with ammonia and nitriding with salt bath. However, these processes are limited with respect to metallurgical control of the nitrided surface, the adhesion of the layer, the environmental aspect and human health, which makes its use critical as companies seek for certification based on ISO 14000.

A solution to avoid these problems was originally proposed by Berghaus in the 1930s, however, at that time, technological limitations prevented the full development of the process. New developments with computerized control and automation of the procedure allowed the development of plasma nitriding and its use on industrial scale.

During the process, a potential difference of 100–1,500 V is imposed between the furnace wall (anode) and the parts (cathode). Under controlled conditions of temperature (between 380 and 600 °C), pressure (between 13 and 1300 Pa) and gas mixture it is possible to generate a bright discharge, plasma, which completely covers the surface of the parts as shown in Fig. 6.2.

Plasma consists of ionized nitrogen and electrons, accelerated within an electric field, according to the following reaction:

$$N_2 \rightarrow 2N^+ + 2e^-$$

The high kinetic energy of the ions collision with the surface of the substrate in the region of the plasma sheath generates heat and does the cleaning, the

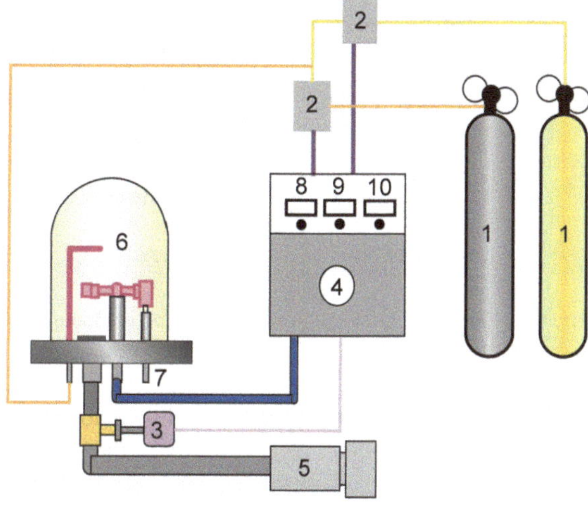

Fig. 6.3 Diagram of the nitriding reactor—Nitrogen and Hydrogencylinders (1); Flow meters (2); Pressure gas system (3); Voltage source DC (4); Vacuum pump (5); Reactor (6); Thermocouple (7); Nitrogen flow (8); Hydrogen flow (9) and Pressure (10)

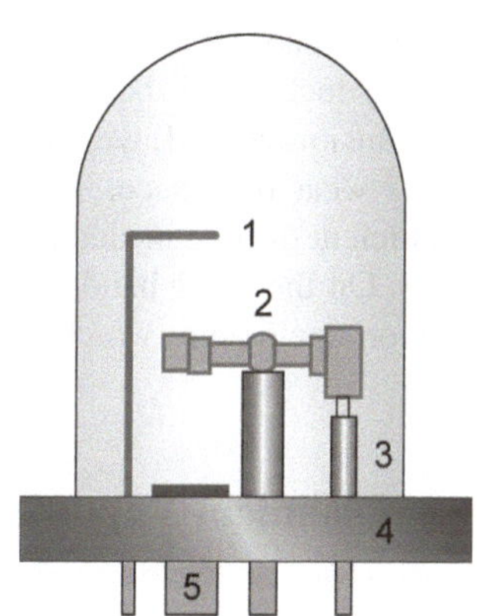

Fig. 6.4 Nitriding reactor—Gas inlet system (1); Samples (cathode) (2); Thermocouple (3); Base of the reactor (anode) (4) and Vacuum system (5)

depassivation and the activation of the surface. Therewith, reactions on the surface and diffusion of nitrogen on the parts are produced. A schematic representation of the plasma reactor used in nitriding is shown in Figs. 6.3 and 6.4.

The biggest advantage of the plasma nitriding is the possibility of controlling the metallurgy of the nitride layer [4]. For single steel, this process allows to vary the type of nitride formed in the layer of compounds and even prevent the formation of this layer. To do so, the following must be carefully controlled: the composition of the gas mixture, the nitriding temperature and time, as shown in Fig. 6.5. It is noteworthy that the chemical composition of the substrate plays an important role on the metallurgy of the nitrided surface [5]. The depth of the nitrided layer is controlled by the temperature and the time of the process. Nitriding without layer of compounds is performed by using a low potential of nitrogen and/or short nitriding time. By increasing the potential of nitrogen and adding methane, a layer of compounds is preferably formed with ε-$Fe_{2\text{-}3}N$ nitride.

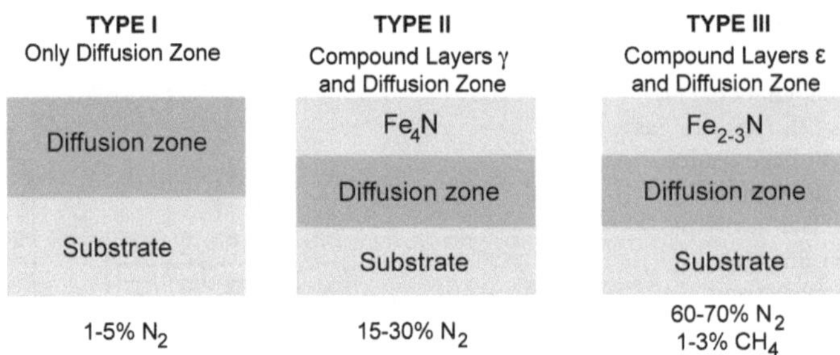

Fig. 6.5 Composition of the several layers of compounds which may be formed during the plasma nitriding

The control of the nitrided layer allows adapting it to its use as follows. The formation of the white layer improves the fatigue resistance due to residual compressive stress and improves microhardness up to values of approximately 1,500 HV. However, very thick white layers tend to stand out creating micro-spallings and damaging the fatigue resistance of the component.

The ductility of the nitrided layer is considerably greater if it is composed only of diffusion zone. Layers of compounds containing only one type of nitride have better wear resistance. The presence of the white layer reduces the dry friction coefficient of the nitrided material, which results in a higher resistance to adhesive wear. On the other hand, nitrided layers formed only by diffusion zone have high friction coefficient, which causes intense adhesion on the surface of the material.

6.3 Chemical Vapor Deposition

The CVD technique originated in the Middle Ages, once we consider ceramic or porcelain vitrification as a primitive form of CVD. However, the first experiences of using CVD to deposit hard materials on steel were performed in 1953/54 by Munster and Ruppel in Germany.

CVD can be defined as a synthesis method in which the constituents of the vapor phase react to form a solid layer on a surface. Therefore, the knowledge about the type of chemical reaction and its extension are essential for the CVD process. Moreover, the effects of several variables of the process such as temperature, pressure, concentration and flow must be considered.

Chemical reactions in CVD systems occur in low-pressure reactors, where the diffusion of an inert gas is inversely proportional to pressure. In that case, as the pressure used is lower than the atmospheric pressure (760 Torr to 0.5–1 Torr), the diffusion is favored and the actual effect is an increase in the gas phase transfer of reactants to and through products of the surface and the determining factor of the velocity is the reaction on the surface. The results are a well adherent layer with good uniformity, good structural homogeneity and good coverage.

A variation in this process improves preparation of the surface "in situ" with plasma by using parts as negative potential. In this case, the heated part is cleaned once more by cathode disintegration. This variation of the process also increases the release force or the ability to uniformly settle in the contour and recessed areas.

The CVD process of coating tools for abrasion resistance is usually conducted at temperatures above 1,000 °C. At these temperatures the parts undergo a self-cleaning treatment, which can be improved by the cathode disintegration, generally with an excellent adhesion layer. This technique is currently widely used for coating of hard metal tools, which are not greatly affected by the high temperature of the process, i.e. they do not require a new quenching, as steels that lose their mechanical strength do.

However, even hard metal tools suffer certain reactions with the vapor phase constituents that may cause some embrittlement, such as a reduction of about 40 % of the flexural strength of certain inserts. This embrittlement is caused by the formation of ε-$Fe_{2\text{-}3}N$ phase, brittle, which can not be prevented and restrict the process for application in hard metal (for example, there has been no success in intermittent cutting operations). As temperatures of 1,000 °C are higher than the tempering temperature of any high speed steel or tool steel, these should be quenched, tempered and grounded again due to distortion problems.

6.4 Physical Vapor Deposition

Temperature restrictions in the CVD process led to researches and new development that culminated with the PVD process, PVD, which uses low temperatures. The coating process can be executed below the tempering temperature of the high speed steel and other steel tools.

The process was accidentally discovered by Faraday, who observed the deposition of particles on the inner surface of incandescent bulbs deriving from the explosion of filaments.

In general PVD is a process in which a reactive metal component in solid state, for example titanium, replaces the CVD gaseous medium of reaction, as shown in Fig. 6.6.

Cutting tools coated with TiN by PVD were introduced in Japan and Europe in the late 1970s [6]. Historically speaking, these were relatively expensive processes, given the low rates of deposition of the coating and the investment required to build the vacuum chambers. Electrodeposition and C.V.D. were commonly less expensive.

More recently, with the development of techniques that allow the deposition of coatings at higher rates and the increased demand for materials and coatings for high performance, the use of PVD increased considerably.

Coatings deposited by PVD processes show performances that are similar to coatings applied by CVD in most applications, including coating of hard metal tools used in intermittent cut, considering that the CVD process often reduces the bending resistance of the hard metal.

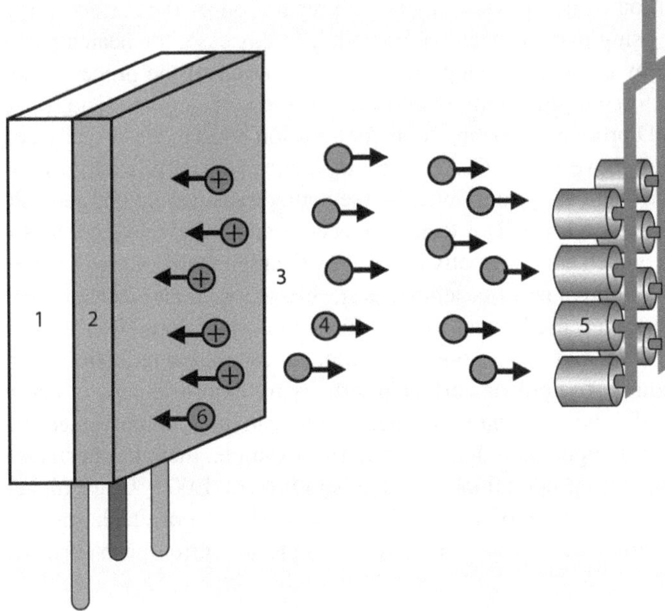

Fig. 6.6 Discharge in the "magnetron"—Magnetron (*1*); Ti target (*2*); Discharge (*3*); Atoms of the target (*4*); Parts to be coated (*5*) and Argon ions (*6*)

PVD is extremely versatile and allows a wide number of combinations between coating materials and the substrate [2]. Metals, alloys, ceramic, intermetallic and some types of polymers can be deposited on the surface of any material (such as metals, ceramics, plastics and paper), provided that they are subjected to stable operating temperatures into the vacuum.

Three different PVD processes are used to sublimate the coating material, as follows.

PVD consists of a great variety of techniques, which can be divided basically into three groups: deposition by vacuum evaporation, cathode sputtering and ion plating. Each of these coating sublimation techniques will be specified below.

6.4.1 Deposition by Vacuum Evaporation

Evaporation is one of the oldest deposition processes into the vacuum. Relatively simple and inexpensive, it provides high deposition rates and is used especially to obtain coatings with greater layer thicknesses [2].

During the deposition by evaporation, the coating material is evaporated by heating the material source at a given temperature (commonly between 1,000 and 2,000 °C) into the vacuum in a way that the vapor pressure exceeds the chamber atmospheric pressure considerably, producing sufficient vapor for the condensation.

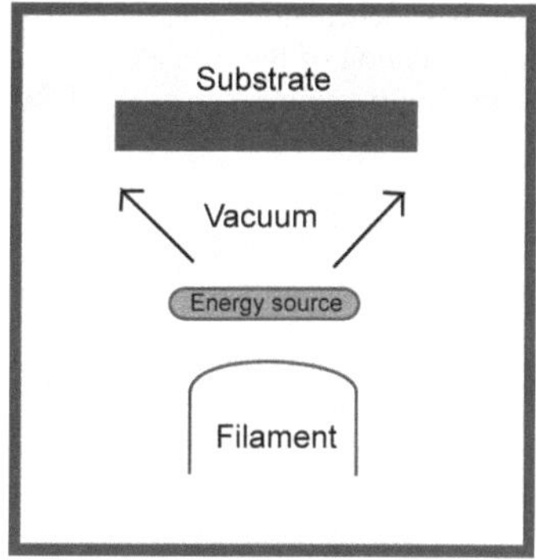

Fig. 6.7 Schematic representation of the evaporation process into the vacuum

The coating is released from the surface of the material source into the substrate, travelling a distance which may vary between 150 and 450 mm. A schematic representation of the process is shown in Fig. 6.7.

The heating of the material source may be accomplished by induction, resistance, electron beam, laser, among others.

The process has low kinetic energy, which reflects a low adhesion of the coating. Furthermore, in order to deposit thicker films, it is necessary to heat the substrate at higher temperatures of around 1,600 °C [2]. Another limitation of this process is the fact that the substrate must be moved to obtain uniformity in the coating deposited.

In some cases, plasma is used in the process to increase the reaction, generate ions and accelerate them into the substrate. This process is called Activated Reactive Evaporation (ARE), which is recognized for providing better coating adhesion, better mechanical properties and high deposition rates, when compared to conventional evaporation deposition processes.

6.4.2 Cathode Sputtering

In this process, positively charged ions are accelerated towards the material to be coated by electrical discharge heating, usually positive ions of an inert or reactive gas.

Argon is the most widely used gas due to its compatibility with the main engineering materials and its low cost. Krypton and xenon are used in some specific applications.

This ion bombardment sublimates them from the coating source and deposits them on the parts with high energy in the form of a pure and highly adhesive layer. In the process, the coating material, also called target or cathode, and the substrate to be coated are placed in a vacuum chamber with a pressure in the range from 0.013 to 13 mPa (10^{-1} a 10^{-4} mtorr). The material is sublimed and ejected in the atomic form from the target. The substrate is placed in front of the target to intercept the flow of ejected atoms. Figure 6.8 shows a schematic representation of the cathode sputtering process.

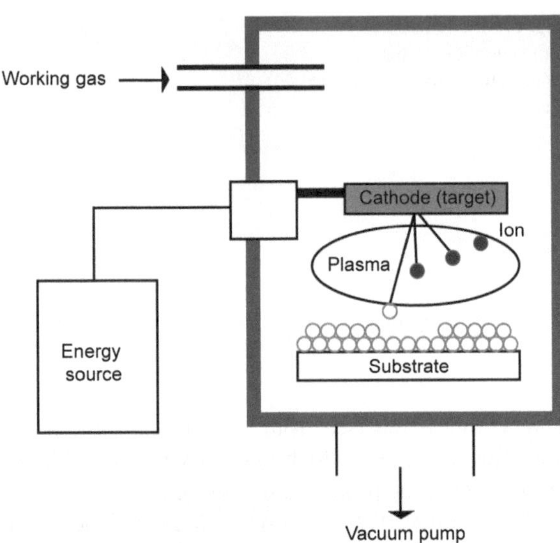

Fig. 6.8 Schematic representation of the cathode sputtering process

It is estimated that only 1 % of the energy reflected from bombardment in the target surface produces ejected atoms, while 75 % cause the heating of the target and the remainder is associated with the emission of secondary electrons.

This process is suitable for uniform coating deposition on large and flat surfaces. The layer thicknesses obtained through cathode sputtering varies between 0.02 and 10 μm [2].

In a conventional process, the discharge generated spreads throughout the chamber, as shown in Fig. 6.9a. Due to low pressures, the average free path of target's ejected atoms is short—about 2.0 mm. They reach the substrate with low energy, impairing the adhesion and density of the deposited film. In addition, high-energy electrons continuously bombard the substrate, raising its temperature and consequently restricting this process to substrates that can withstand such temperature rise without changing.

In order to increase the versatility and the field of application of the cathode sputtering, a process called Magnetron Sputtering was developed. Magnets placed in the vicinity of the target produce flux lines forming a closed path in front of the target, as shown in Fig. 6.9b. Thus, "electron traps" are formed. The magnetic field exerts some force on the secondary electrons emitted by the cathode, which causes them to follow the flux lines path. Due to the formation of these "electron traps", the plasma is confined around the cathode surface, which increases the process efficiency.

The temperature of the parts to be coated can be kept below the tempering temperature of, for example, high speed steel and, therefore, they can be finished before the coating process without any dimensional change. The cathode sputtering is always possible to be made and several materials can be deposited. The disadvantage of the process is the low deposition rate and some difficulties in obtaining uniform coatings on parts with recess.

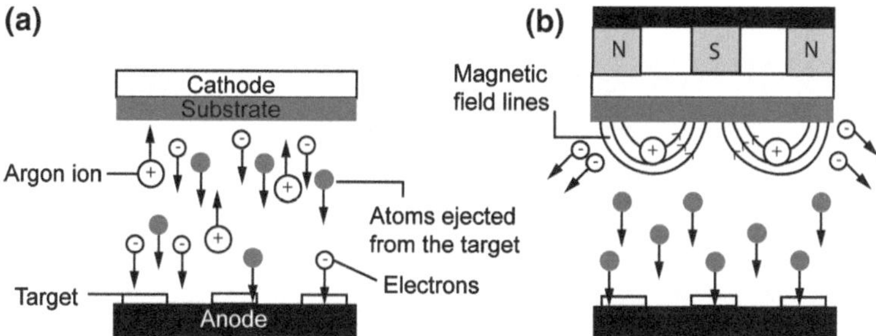

Fig. 6.9 **a** Conventional cathode sputtering and **b** Magnetron sputtering

6.4.3 Ion Plating

The third technique used in the PVD technology is the plasma-assisted evaporation technique, better known as Ion Plating. Contrary to the previous process of cathode sputtering, in this case parts are subjected to ion bombardment before and during the deposition process. In one way, the evaporation technique made by plasma combines the advantages of the cathode sputtering (good appearance) with the high deposition rate of the vacuum deposition process. Normally, a good uniformity of coatings, even in parts with recesses, can be obtained, which currently consolidates the technique as the most widely used.

The term "Ion Plating" was coined in 1961 by D.M. Mattox, although the process had been patented by Berghaus in 1936. Mattox used the term "Ion Plating" to describe all the techniques of processes in which the growth of thin layers is performed through particle bombardment with high energy to obtain better adhesion of the layer and special properties. Currently, the term "Ion Plating" is often used for the PVD technique for layers that are resistant to friction in high speed steel tools.

In a conventional process of "Ion Plating", the parts are held at high voltage in direct current (normally 1–5 kV) in an inert gas atmosphere, e.g. argon, at about 1×10^{-2} mbar. Positive ions heated by the discharge are accelerated by an electric field created by the cathode fall and, as a result, high energy ions start to bombard the surface of the parts. Thereafter the parts are continually cleaned before and during the deposition process. The coating material evaporates due to the heated discharge at low pressure where it is dispersed and partially ionized by multiple collisions.

Ions of the coating material in the heated discharge surround the parts and are accelerated along the surface through the cathode dark field. The ions are projected over the surface of the parts with high energy on account of the cathode dark field presenting higher field gradient. In many cases, however, the predominant process is that of high pressure neutral atoms condensation of the coating material. Many neutral atoms have high energy because they have been ionized by a certain period of time or have been driven by accelerated ions.

Thus, two different phenomena occur simultaneously on the surface of the parts treated: a deposition of the coating material (neutral atoms and ions), and a

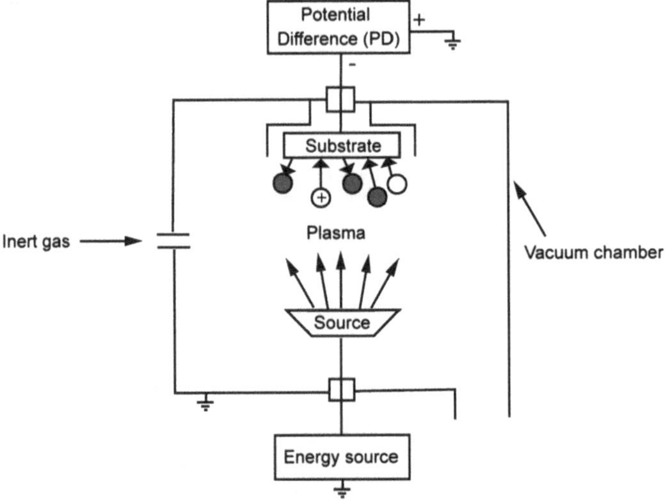

Fig. 6.10 Schematic representation of the ion plating process

cathode sputtering made by argon and coating material (ions and neutral atoms of high energy). Therefore, the effective deposition rate is determined by relative rate of these two phenomena. The layer formation and its growth will only occur if the deposition rate of the coating material overcomes that of the cathode sputtering. In general, all coatings deposited through "Ion Plating" are characterized by:

- Strong layer adhesion, even in the case of materials that do not normally form adherent interfaces;
- High uniformity of coating on materials with recesses;
- High layer density and extremely fine grain structure.

Treatments in which occurs formation of two layers, called PAPVD have given excellent performance to the tools. In such cases the plasma nitriding is performed in order to form a substrate of higher hardness in the tool, over which it is applied a thin TiN layer (5 μm) through PVD by using reactors that are similar to those of plasma nitriding and obtaining microhardness of about 2,500 HV. Figure 6.10 shows a schematic representation of the ion plating process.

6.5 Modern Coatings

6.5.1 Multilayer

The development of the PVD process allows tools to be coated with multilayers and enables one to control the sequence of components and the layers thicknesses. The multilayer structure alternates between high and low stress, resulting in high hardness,

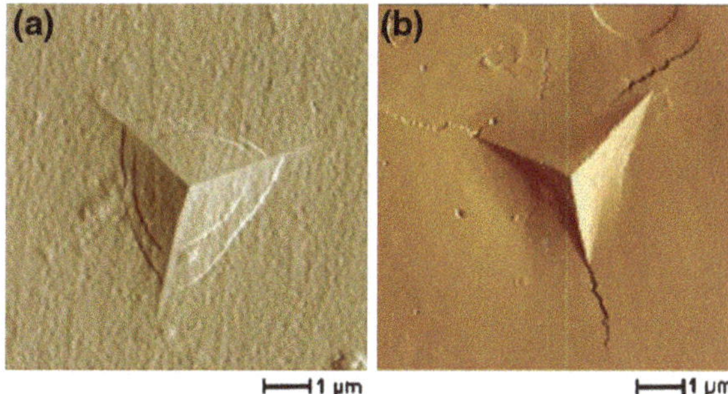

Fig. 6.11 Propagation of cracks in **a** multilayer and **b** monolayer films

good adhesion, good toughness and little fatigue, which increases the wear resistance and reduces the presence of cracks when compared to monolayer. Figure 6.11 shows a comparison of crack propagation between monolayer and multilayer.

Several classes of multilayers have emerged in the last decade, which makes the application of these coatings very interesting for tribology. An example of a new class of multilayers is the hardness superlattice coating.

Hardness superlattice is a coating that present layers with thickness shorter than 10 nm and high mechanical resistance (>50 GPa) [7]. Several multilayer systems have been developed, including TiN/TiAlN, TiN/CrN, TiAlN/CrN, TiN/ZrN and CrN/AlN. The multilayer of titanium nitride and titanium aluminum (TiN/TiAlN), for example, can be applied over substrates of hard metal and high speed steel. This multilayer presents hardness of 3,300 HV 0.05 and excellent wear resistance at high temperatures.

6.5.2 DLC: Diamond-Like Carbon

In the last decade, the search for carbon-based coatings with lubricant characteristics emerged as an important solution to the reducing of friction and wear in machinery and equipment [8].

Recent progress on coating technology already allows the deposition of films with properties that could not be achieved a decade ago [9].

Some examples are the multilayer coatings, nanocomposites and nanostructured, gradients and metastable with extreme mechanical and chemical properties.

Carbon is known as a unique element that offers various types of structures due to its chemical bonds, as well as fullerenes, nanotubes, two and three-dimensional films among others, used by various components and applications [10].

Figure 6.12 shows an example of two different carbon structures which, due to their structures, present significant differences in their physical–chemical properties.

Graphite is a lamellar structure of carbon widely used as solid lubricant. Due to low bonding energy, Van der Waals forces, crystallographic planes slip allows to reduce friction of tribological systems. In turn, diamond presents a three-dimensional structure in which each carbon atom is chemically bonded to four other

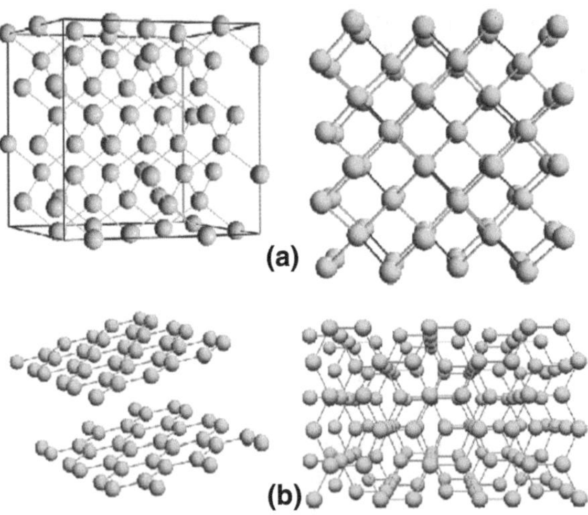

Fig. 6.12 Crystalline Structures of carbon: Diamond lattice (**a**) and Graphite lattice (**b**)

carbon atoms. The result is a crystalline structure with high bonding energy giving diamond high hardness and wear resistance.

Although these compounds have essentially the same constituents, their properties and applications are completely different.

Carbon-based coatings, also called Diamond-like Carbon coatings (DLC), are currently supplied on industrial scale and have been used to reduce friction and wear, even under dry application [10].

Carbon's ability to form different compounds is related to the types of hybrid bondings (sp^3, sp^2 and sp^1) this element can create.

In sp^3 configuration, the four electrons from the valence shell of a carbon atom tend to form strong tetrahedral bonds, as in diamond and DLC films.

In graphite, there are sp^2 configurations, in which three out of the four valence shell electrons form trigonal bonds in a plane. Figure 6.13 shows the types of carbon bonding.

Carbon-based coatings vary greatly in their properties according to the proportion of sp^2 and sp^3 bondings. Likewise, the presence of hydrogen and of alloying elements plays a key role in the properties of such coatings.

The tribological behavior of carbon films depends also on the chemical and structural characteristics obtained through different techniques for coating deposition [10].

A ternary phase diagram was proposed by Silva et al. [11] in which the DLC films are arranged according to their hydrogen content and the ratio of sp^2 and sp^3 bondings.

The increase in the sp^3 hybrid bonding portions gives DLC coatings characteristics similar to those of the diamond (sp^3 hybrid bondings >85 %). Such coatings are called tetragonal amorphous carbon or ta-C.

In turn, the increase in the sp^2 hybrid bonding portions gives coatings characteristics similar to graphite, and is called the amorphous carbon or a-C.

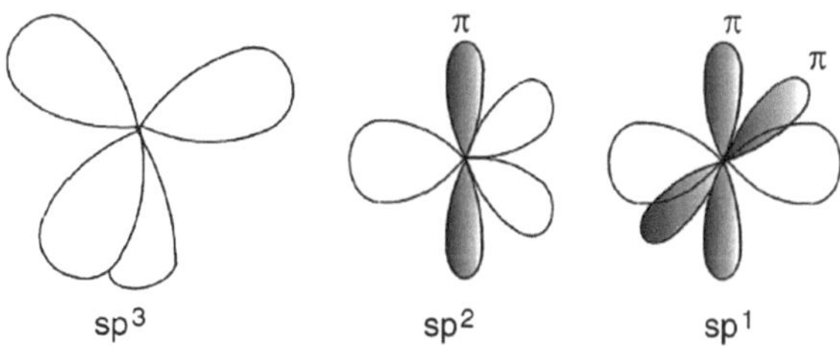

Fig. 6.13 Representation of sp^3, sp^2 and sp^1 bondings

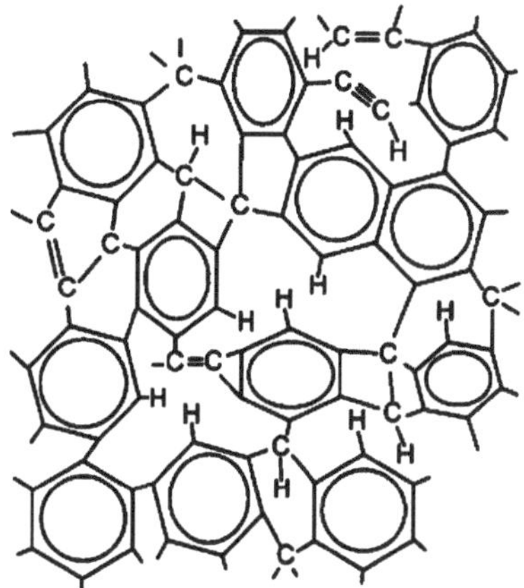

Fig. 6.14 Chemical structure of DLC

The incorporation of varying hydrogen content (hydrogenation) in the coating allows the variation of chemical structures and facilitates coating manufacturing. Such coatings are more common and are called hydrogenated amorphous carbon or hydrogenated amorphous tetragonal carbon, a-C:H or ta-C:H, respectively [10, 12].

The films that are not hydrogenated present hydrogen content bellow 1 %. In turn, hydrogenated films may present up to 50 % of hydrogen [13]. Figure 6.14 presents the chemical structure of a DLC coating of type a-C:H.

Robertson [12] defined DLC as "a metastable form of amorphous carbon with significant proportions of sp^3 bondings".

In summary, DLC coatings are thin films that can vary greatly in composition and structures. The structure of the film and its properties are also determined by the hydrogen content and the ratio of sp^2 and sp^3 hybrid bondings that compose their structures. [11].

Donnet [14] proposed the modification of certain properties of the DLC by adding alloying elements such as: B, Si, Cr, F, W and N, and other metals.

A graphical presentation of the different types of DLCs and their use are shown in Fig. 6.15.

According to Edermir [10] the addition of Si to the DLC can improve the friction and wear properties and make these films highly insensitive to changes in humidity and temperature. The addition of Cr and/or W to the DLC increases the resistance to adhesion under conditions of lubricated slip. On the other hand, the DLC films that contain nitrogen are currently used in applications involving computer hard drives.

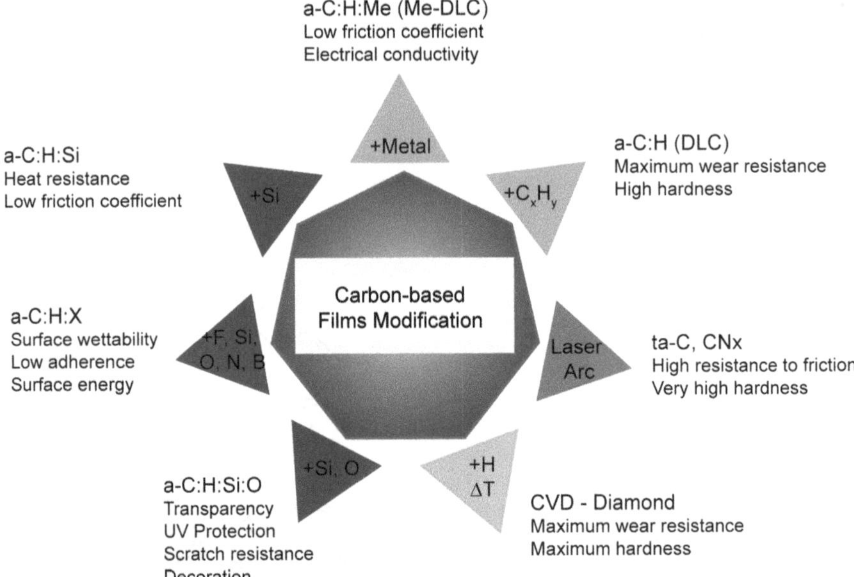

Fig. 6.15 Applications of DLC according to their structures and chemical compositions

All the processes of deposition of DLC occur under non-equilibrium conditions and are characterized by the interaction of energetic specimens of carbon and the substrate surface forming the film [9].

6.5.2.1 Process of DLC Coating Manufacture

DLC coatings can be manufactured through various methods of manufacture. The method of manufacture can be determined considering the use of the coating, whether via laboratory or industrial output [12].

The deposition of DLC coating can be accomplished through different methods at low temperatures and high deposition rates by using processes such as PACVD—Plasma Assisted CVD, IBD—Ion-beam deposition, DC—Direct Discharge current, RF—Radio Frequency, cathode spraying—Sputtering, electron beam evaporation—Ion Plating, and cathodic arc methods.

However, the deposition of tribological coatings and solid lubricants may require deposition temperatures ranging from room temperature to over 1,000 °C. In some of these methods, high temperatures of deposition may cause undesired phase transitions, softening of the substrate and change in shape [9].

A variation in the CVD coating process, also known as PACVD is one of the most important techniques used for deposition of thin films allowing a wide variety of crystalline and amorphous materials.

Figure 6.16 shows a cross section that exposes the CrN and DLC coatings obtained by scanning electron microscopy—SEM.

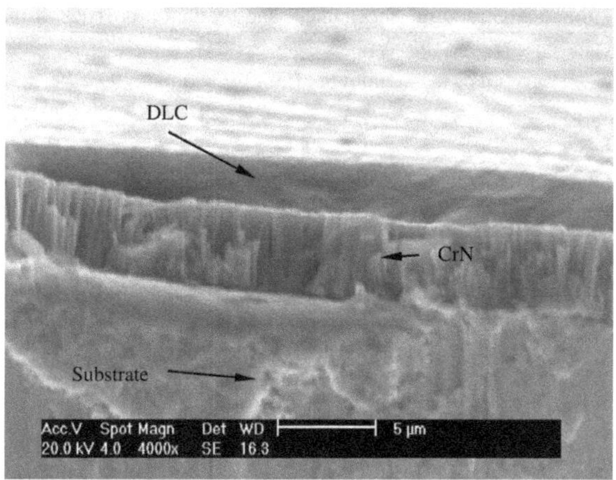

Fig. 6.16 Picture of the cross section obtained by scanning electron microscopy

6.6 Coating Properties

Coatings applied by PVD techniques have made a major impact on industrial processes. Increase in tool's life, reduction of the number of downtime periods, reduction of maintenance costs, productivity gains, best finish of final products and reduction or elimination of lubricating fluids are some of the benefits achieved through the application of these coatings [3].

Some properties of the coatings, such as high hardness—reaching five times that of quenched and tempered high speed steel—, low friction coefficient and high chemical and thermal stability, are the main factors responsible for the benefits mentioned above. Other advantages inherent to the application of coatings by means of PVD are noteworthy, for example, the ability to treat without changing the structure of the material—which is not possible in the CVD process—, the minimal environmental damage, which does not produce waste and therefore represents an alternative to galvanic coating and a possibility of removal followed by reapplication of the layer [3].

The coatings' performance is largely determined by its hardness. Table 6.2 [15] shows the properties of materials commonly used in hard coatings, compared to steel common properties. It is observed that the coating materials are much harder, and present at least twice the modulus of elasticity and have coefficients of thermal expansion between 0.5 and 1.0 of the coefficient of thermal expansion of steel. Melting points are much higher and show high chemical stability.

More recently, in addition to hardness, other properties of the coatings have been extensively studied. High toughness and resistance to corrosion, oxidation and tribo-oxidation are trends in the development of new tribological films [3].

Table 6.3 presents some properties of titanium nitride, chromium nitride and titanium carbonitride.

The microhardness values available in the literature [16–18] differ widely, especially on titanium nitride and chromium nitride. Some of these values are

Table 6.2 Structural, mechanical and thermal properties of materials used in hard coatings [15]

Material	Hardness (GPa)	Young's modulus (GPa)	Thermal conductivity (W/m K)	Coefficient of thermal expansion (10^{-6} K^{-1})	Melting point (°C)
TiN	20	440	29	9.4	2,949
TiC	29	450	18–30	7.4	3,067
VN	15	460	–	8.0	–
VC	29	430	–	–	2,648
NbN	14	480	–	10.1	2,204
NbC	24	580	–	6.6	3,600
CrN	11	400	84	10.3	1,810
WC	21	695	14	4.3	2,776
Al_2O_3	21	400	350	9.0	2,300
SiC	26	480	27	5.3	–
BN	40	660	–	–	–
B_4C	35	440	–	5.0	–
Diamond	90	590	33	0.8	–
Steel tool	8	220	16	13.0	1,500

Table 6.3 Properties of some superficial coatings

Coating	Layer thickness (μm)	Friction coefficient	Maximum operating temperature (°C)
Titanium nitride	1.0–6.0	0.5	550
Chromium nitride	1.0–20.0	0.4	450
Titanium carbonitride	1.0–4.0	0.4	700

Table 6.4 Values of microhardness available in the literature for the coatings of titanium nitride, chromium nitride and titanium carbonitride

Reference	Microhardness [HV 0.05]		
	Titanium nitride	Chromium nitride	Titanium carbonitride
(Nordin; Larsson; Hogmark 1999)	2310 ±380	1,950 ± 160	–
(Su et al. 1997)	2,756	2,065	2,854
(Heinke et al. 1995)	2,000 ± 346	2,306 ± 224	–
(Bender; Mumme 2005)	2,300	1,750	3,000
(Durotin 2004)	2,500	2,300	3,000

presented in Table 6.4. It is believed that such wide variation in the microhardness values may be due to different techniques and process parameters used to obtain these coatings, by forming layers with different characteristics, such as presence phases and preferential orientations, and thus different properties [3].

Titanium nitride is the most widely used coating in industry due to its hardness properties, adhesion layer and resistance to elevated temperature, which makes it a coating of universal use. Furthermore, it is a biocompatible material, whose use is

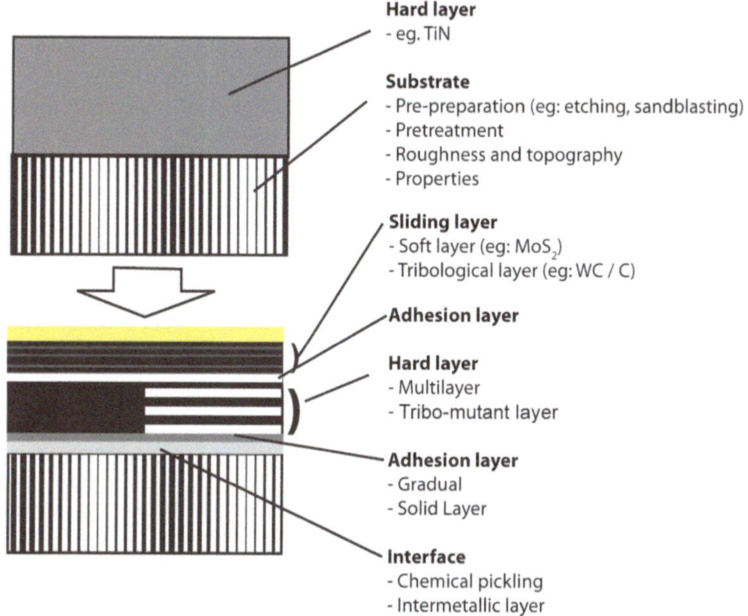

Fig. 6.17 Chematic representation of the two main configurations for the formation of coating layers

approved by the food industries. The TiN obtained by PVD process of low temperature, or ion plating, has presented excellent results when applied on high speed steel tools [2].

Chromium nitride is one of the most widely used coatings on industry to increase the wear resistance of components. This coating is generally applied at relatively low temperatures around 200 °C and has excellent adhesiveness.

Furthermore, the chromium nitride layer can be applied to surfaces with low roughness, which allows the reduction of internal stresses, even for thicker layers.

The titanium carbonitride presents hardness superior to that of TiN and CrN, high abrasion resistance and low coefficient of friction. Tools coated with TiCN present high wear resistance. It is used in machining of low machinability alloys such as hardened steels, and other high abrasion operations.

Recent developments can be divided roughly into two lines of research. The first one, most widespread, studies the systems composed of solid layers, i.e. configurations consisting of substrate and a solid film, uniform and of homogeneous distribution from the beginning to the end of their formation. More traditional coatings such as titanium nitride (TiN) and chromium nitride (CrN) follow that line. The second one studies the systems whose characteristics and properties may vary significantly from the beginning to the end of their deposition. These coatings are called tribo-mutants, which are composed of multilayers with different characteristics.

Figure 6.17 shows a schematic representation of this type of coating. In this example, the outermost layer has a minimum coefficient of friction and low wear resistance, acting as a lubricant. With the progress of wear due to the properties of the second layer, there is a gradual increase in the level of resistance to abrasion and friction.

References

1. Bell T (1991) Surface treatment and coating of PM components. Powder Metall 34:253–254
2. Bhushan B, Gupta BK (1991) Handbook of tribology: materials, coatings and surface treatments. McGraw-Hill, New York
3. Konig RG (2007) Estudo do desgaste de revestimentos em matrizes de recorte a frio de cabeças de parafusos. UFSC
4. Edenhofer B (1974) Physical and metallurgical aspects ionitriding, Heat Treat Met 1:23–28
5. Jones CK et al (1973) Ion nitriding. Heat Treat 73:71–77
6. Wick C, Veilleux RF (1986) Vapour deposition process. In: TOOL and manufacturing engineers handbook:materials, finishing and coating (4th edn) vol 3
7. Musil J (2000) Hard and superhard nanocomposite coating. Surf Coat Technol 125:322–330
8. Silvério M (2010) Comportamento tribológico de revestimento multifuncional CrN-DLC em atmosferas de gases refrigerantes. Dissertação (mestrado), UFSC, Centro Tecnológico
9. Donnet C, Erdemir A (2004) Historical developments and new trends in tribological and solid lubricant coatings. Surf Coat Technol 180–181:76–84
10. Erdemir A (2004) Design criteria for super lubricity in carbon films and related microstructures. Tribol Int 37:577–583
11. Silva SRP, Robertson J, Milne WI, Amaratunga, GAJ (1998) Deposition mechanism of diamond-like carbon, in amorphous carbon: state of the art. World Scientific Publishing, pp 32–45
12. Robertson J (2002) Diamond-like amorphous carbon. Mater Sci Eng R 37:129–281
13. Grill A (1999) Diamond-like carbon: state of the art. Diam Relat Mater 8:428–434
14. Donnet C (1998) Recent progress on the tribology of doped diamond-like and carbon alloy coatings: a review. Surf Coat Technol 100–101:180–186
15. Hultman L, Sundgren JE (2001) In: Bunshah RF (ed) Handbook of hard coatings: deposition technologies, properties and applications. Noyes Publications, Park Ridge, New Jersey, pp 111
16. Nordin M, Larsson M, Hogmark S (1999) Mechanical and tribological properties of multilayered PVD TiN/CrN. Wear [S.I.] 232:221–225
17. Su YL et al (1997) Comparison of tribological behavior of three films TiN, TiCN and CrN—grown by physical vapor deposition. Wear 213:165–174
18. Heinke et al (1995) Evaluation of PVD nitride coatings, using impact, scratch and Rockwell-C adhesion tests. Thin Solid Films, [S.I.] 270:431–438

Index

J. P. Davim (ed.), *Tribology in Manufacturing Technology*, Materials Forming, Machining and Tribology, DOI: 10.1007/978-3-642-31683-8,